Handbook
of
Digital Electronics

Handbook
of
Digital Electronics

JOHN D. LENK

Consulting Technical Writer

PRENTICE-HALL, INC., Englewood Cliffs, New Jersey 07632

Library of Congress Cataloging in Publication Data

Lenk, John D
  Handbook of digital electronics.

  Includes index.
  1. Digital electronics. I. Title.
TK7868.D5L44   621.3815   80-19647
ISBN 0-13-377184-9

Editorial/production supervision and interior design
    by Barbara A. Cassel
Manufacturing Buyer: Joyce Levatino

Printed in the United States of America

10 9 8 7 6 5 4 3 2 1

PRENTICE-HALL INTERNATIONAL, INC., *London*
PRENTICE-HALL OF AUSTRALIA PTY. LIMITED, *Sydney*
PRENTICE-HALL OF CANADA, LTD., *Toronto*
PRENTICE-HALL OF INDIA PRIVATE LIMITED, *New Delhi*
PRENTICE-HALL OF JAPAN, INC., *Tokyo*
PRENTICE-HALL OF SOUTHEAST ASIA PTE. LTD., *Singapore*
WHITEHALL BOOKS LIMITED, *Wellington, New Zealand*

To Irene, the "Sandpiper Lady"
and Mr. Lamb, the "Magic Lamb"

Contents

CHAPTER 5
Testing and Troubleshooting Digital Equipment   299

Preface

This book is a "crash course" in digital electronics, written with five classes of readers in mind. First, there is the engineer who must design with (and program) digital devices. Next is the technician who must service digital electronic equipment. Then there is the programmer/analyst who wants to relate programs and software to digital hardware. Finally, but not of least importance, are students and hobbyists who want an introduction to digital electronics, as well as information they can put to immediate use in experiments and projects.

The various classes of readers start from a different learning point. Obviously, the engineers and technicians understand electronics, but they may have little knowledge of digital circuits or no knowledge of programmed devices. Programmers may be expert in computer language and systems, but not know anything of digital devices, particularly microprocessor-based devices. The student/hobbyist may have only an elementary knowledge of electronics and no understanding of digital or programmed equipment.

   This book bridges the gaps by bringing all these readers up to the same point of understanding. This is done in a unique manner, following the tradition of the author's best-sellers. The descriptions of how digital electronic devices operate are technically complete (to satisfy the technicians and engineers) but are written in simple, nontechnical terms wherever possible (for the benefit of programmers/students/hobbyists).

   Another major purpose of this book is to provide a simplified system of test and troubleshooting for digital electronic equipment. While attempts have been made to do so, it is not only impractical, but virtually impossible, to cover troubleshooting for every type of digital device in use. Books that attempt to do so are soon made obsolete by rapid technological advances. To overcome this problem, we concentrate on basic approaches to digital test and troubleshooting. These approaches can be applied to any digital electronic system now being developed, to those that will be manufactured in the future, and to digital equipment now in use.

   The test and troubleshooting approaches found here are based on the techniques found in the author's best-selling books: *Handbook of Practical Solid-State Troubleshooting, Handbook of Basic Electronic Troubleshooting, Handbook of Simplified Television Service, Handbook of Practical CB Service,* and *Handbook of Practical Microcomputer Troubleshooting.*

   Chapter 1 of this book provides a quick review of the number systems and alphanumeric codes used in digital electronics. Chapter 1 also describes the relationships between the number systems and the electrical signals or pulses that control digital electronic equipment.

   Chapter 2 is an introduction to digital logic, which includes the basics of digital logic circuits, the logic symbols in common use throughout the industry, the basic principles of logic equations and logical (Boolean) algebra (including mapping), and corresponding hardware functions. This chapter also provides a description of both combinational and sequential networks, including gates, decoders, encoders, code converters, data distributors and selectors, multiplexers, parity and comparison circuits, flip-flops, latches, counters, and registers. Chapter 2 summarizes the entire subject of digital logic, on the assumption that many readers will be students who are not familar with these basics on which digital equipment operates.

   Chapter 3 describes a cross section of digital circuits, particularly integrated circuits or ICs, in common use. The chapter starts with a discussion of the available digital IC types or forms. Then the chapter proceeds with descriptions of such digital circuits as arithmetic logic units (ALUs), analog/digital (A/D) and digital/analog (D/A) converters, digital readouts and displays, digital interface circuits, and digital memory systems.

Chapter 4 describes all elements of a typical digital computer system. Microprocessor-based computer systems are stressed. Such subjects as microcomputer terms, elements, circuit functions, hardware, memories, input/output, basic programs, machine and assembly language, peripherals, terminals, floppy disks, printers, and digital communications are covered.

Chapter 5 is devoted entirely to descriptions of the procedures and equipment required for test and troubleshooting of digital electronic equipment. Such subjects as digital test equipment, basic digital measurements, digital equipment service literature, digital IC test and troubleshooting, and examples of digital troubleshooting are covered. The discussions in this chapter stress the way each type of test equipment is used in digital electronics test and troubleshooting, and what signals or characteristics are to be expected from each. Specific techniques for using the test equipment in troubleshooting situations are discussed throughout the final chapter.

Many professionals have contributed their talent and knowledge to the preparation of this book. The author gratefully acknowledges that the tremendous effort required to make this book such a comprehensive work is impossible for one person, and he wishes to thank all who have contributed directly and indirectly.

Special thanks are due to the following: Walter Dennen of RCA Corporation, Solid State Division; Lothar Stern of Motorola Semiconductor Products, Inc.; Louis Frenzel and Coy Clement of Heath Company; Jim Huffhines of Texas Instruments Incorporated; Ross Snyder, Art Pettis, and Larry Shergalis of Hewlett-Packard; and Joseph A. Labok of Los Angeles Valley College.

John D. Lenk

Handbook
of
Digital Electronics

CHAPTER

1

Digital Number Systems and Alphanumeric Codes

Although the outside world generally uses the familiar decimal number system, the circuits found in digital electronics most often use some form of the *binary* number system. This is because binary numbers are compatible with the *electrical pulses* used in digital or logic systems. (When the word "logic" is used, it generally implies "digital logic." The basics of digital logic are discussed fully in Chapter 2). Binary numbers use only two digits, 0 or 1, instead of the 10 digits found in the decimal system. In binary, the 0 can be represented by the absence of a pulse, with the 1 being represented by the presence of a pulse, (or vice versa in some systems). The pulses, typically about 5 V in amplitude, and a few microseconds (μs) or nanoseconds (ns) in duration, can be positive or negative without affecting the binary number system, as long as *only two states exist*. In any event, to understand the language of digital electronics, you must examine number systems in general, and the binary number system in particular.

When the binary number system is used in computer and microcomputer systems, electrical pulses arranged in binary form are often referred to as *machine language*. This machine language, although quite compatible

1

with electronic circuits, is cumbersome when the values are beyond a few digits. For that reason, most computer systems use some other form of number system for assembly of computer programs. Such systems (sometimes referred to as *assembly language*) are essentially shorthand versions of the binary system. The most common shorthand number systems used to enter and read out programs in computers and other digital electronic devices are the *octal, hexadecimal* (or *hex*), *binary-coded decimal* (or BCD), and alphanumeric systems or codes. We discuss each of these systems throughout this chapter. If you are already familiar with digital number systems, you can skip directly to Chapter 2.

1-1 BINARY NUMBER SYSTEM

In all number systems, digits are assigned *positional values* or weights, so that numbers can be written to express all quantities, no matter how large or small. The real value of a digit depends on its position in the number. In binary, the positional values of the digits increase from right to left as in the familiar decimal system. However, with binary, the increase of value is in ascending powers of 2. Thus, the digit at the extreme right (the least significant digit, or LSD) has a value of 2^0, or decimal 1. A digit in the position immediately to the left has a value of 2^1, or decimal 2. The next digit to the left is equivalent to decimal 4; and so on. This can be shown as follows:

2^7	2^6	2^5	2^4	2^3	2^2	2^1	2^0	Binary
128	64	32	16	8	4	2	1	Decimal
MSD							LSD	

When all of these decimal values are added, the total is 255. Thus, an eight-digit binary number can represent decimal numbers from 0 to 255.

1-1-1 Words, bits, bytes, and nibbles

Note in the table above that the extreme left-hand digit (the most significant digit, or MSD) has a value of 128. Also note that this combination of eight binary digits is commonly used in computers, particularly microcomputers, to form a binary *word*. In digital work, binary digits are referred to as *bits* (a contraction of *bi*nary digi*ts*). The LSD bit is 0, or b_0, whereas the MSD bit is 7 or b_7. When eight binary digits are used to form words, the arrangement is known as an *8-bit system*. When all 8 bits are used at once, the combination is referred to as a *byte*. Any number less than 8 bits is called a *nibble,* although in an 8-bit system, a nibble is generally considered to be 4 bits. Similarly, some digital systems use 4-bit or 16-bit binary words or bytes. In fact, many of the newer microcomputer systems are designed for 16-bit words but will also accept 8-bit words so as to be compatible with older systems.

1-1-2 Binary number values

In binary, if the digit is zero, its value is zero. If the digit is 1, its value is determined by its position from the right. For example, to represent the decimal number 38 in binary form, the following combination of 0s and 1s can be used:

b_7	b_6	b_5	b_4	b_3	b_2	b_1	b_0	(bit number)
0	0	1	0	0	1	1	0	(binary number)
128	64	32	16	8	4	2	1	(value)

$0 + 0 + 32 + 0 + 0 + 4 + 2 + 0 = 38$ (decimal)

which means that 00100110 in 8-bit binary form = 38 in decimal.

Note that is not always necessary to use the full 8 bits to write a binary number. For example, decimal 15 requires only four binary digits, 1111, and need not be written as 00001111 on paper. However, in an 8-bit digital circuit, all 8 bits must be included, by adding the 0s.

1-1-3 Number identification

There are many schemes used to identify binary numbers and the other number systems used in digital electronics. In one system, where binary and decimal numbers are mixed, a number subscript (such as 2 for binary, 10 for decimal) is added to minimize confusion. For example, 101_{10} is decimal "one hundred and one," whereas 101_2 is binary "one zero one" or decimal 5_{10}. In other digital equipment literature, a letter is added to the number (D or decimal, B or binary) for identification. Two other commonly used digital number systems, octal and hex (described in the following paragraphs), are identified by number subscripts (8 or octal, 16 for hex) in some literature.

1-1-4 Counting in binary

The counting method for binary numbers is the same as for the decimal system except that we use only two digits, 0 and 1. Thus, starting with 0, if we add one count, we get 1. Adding another count, we go back to 0 again, and a carry is added to the next position at the left, producing 10 (decimal 2). Adding another count produces 11 (decimal 3); and if another count is added, we get 100 (decimal 4). The next count produces 101 (decimal 5); and so on.

1-1-5 Binary fractions

Binary fractions use essentially the same rules that apply to decimal fractions. A binary point (equivalent to the decimal point) is used to separate the integers (whole numbers) from the fraction. The value of the digit to the right of the binary point is 2^{-1}, or the fraction $1/2$ (decimal 0.5). The next position to the right has a value of 2^{-2}, or the fraction $1/4$ (decimal

3

0.25); the next position to the right has a value of 2^{-3}, or the fraction $1/8$ (decimal 0.125); and so on.

Using this system, the mixed binary number 101.101 may be interpreted as $4 + 0 + 1 \cdot 0.5 + 0 + 0.125$, or 5.625 in decimal.

1-1-6   Signed binary numbers

Thus far, we have assumed that all binary numbers are positive. This is the case for *true, natural,* or *pure binary* numbers. However, the *signed binary* system may be used to represent both positive and negative numbers. With signed binary, the extreme left bit of a binary number or word is used to indicate the *sign* of the number, and the remaining bits give the value. The number is positive (+) if the leftmost bit is 0; the number is negative (−) when the leftmost bit is 1.

Using signed binary, when the sign bit is changed from 0 to 1, the whole number is made negative, but the value is not changed. For example:

$$01101011 = +107 \quad \text{(decimal)}$$
$$11101011 = -107 \quad \text{(decimal)}$$

1-1-7   Conversion from binary to decimal numbers

To convert binary numbers to decimal numbers, multiply each binary digit by its positional value, and add all the products, as is done in Sec. 1-1-2. There are other methods for conversion, but they are generally not any simpler.

1-1-8   Conversion from decimal to binary numbers

Again, there are several ways to convert from decimal to binary. The most obvious system is to make up a chart showing the powers of 2, as is done in Sec. 1-1-2, and then count the necessary number of binary 1s and 0s to make up the desired decimal number.

For example, assume that the decimal number 33 is to be converted. The number 33 is more than 32 (sixth position from the right) but less than 64 (seventh position from the right). This means that you need a combination of six digits (1s or 0s, probably both).

Start with the sixth position, or 32. Since you want number 32, write a 1 in the sixth position. Then move to the fifth position, or 16. Thirty-two plus 16 is greater than the desired 33, so use a zero for the fifth position. The fourth position is 8, and 8 plus 32 is more than 33, so you use a 0 for the fourth position. The same is true of the third position, or 4, and the second position, or 2, so both these positions use a 0. The first (right-hand or LSD) position is 1, and 1 plus the sixth position (or 32) makes the desired 33, so both of these positions require a 1. Thus, the pure binary equivalent of decimal 33 is 100001. If the 8-bit system is used, the binary number will read 00100001 for a decimal 33.

1-1-9 Binary addition

As in the case of decimal addition, binary addition is essentially a counting process. However, there are only four simple rules for binary addition:

$$0 + 0 = 0 \quad \text{with no carry}$$
$$0 + 1 = 1 \quad \text{with no carry}$$
$$1 + 0 = 1 \quad \text{with no carry}$$
$$1 + 1 = 0 \quad \text{with a carry of 1}$$

For example, to add binary numbers 11101 and 1011:

Binary						*Decimal Equivalent*	
+1	+1	+1	+1			(carry)	
	1	1	1	0	1	(augend)	29
+			1	0	1	1	(addend)	+ 11
1	0	1	0	0	0		40

The same rules apply to the addition of binary fractions and the addition of binary mixed numbers. However, great care must be taken so that the binary points of both the augend and the addend are lined up one below the other.

1-1-10 Binary subtraction

There are also four simple rules for binary subtraction:

$$0 - 0 = 0 \quad \text{with no borrow}$$
$$1 - 1 = 0 \quad \text{with no borrow}$$
$$1 - 0 = 1 \quad \text{with no borrow}$$
$$0 - 1 = 1 \quad \text{with a borrow of 1}$$

For example, to subtract binary 0111 from binary 1001:

Binary					*Decimal equivalent*	
−1	−1			(borrow)		
1	0	0	1	(minuend)	9
−	0	1	1	1	(subtrahend)	− 7
0	0	1	0	(difference)	2

Note that when the first borrow is subtracted from the 0 of the minuend, it leaves a 1 with a borrow from the digit at the left. When the 1 of the subtrahend is subtracted from the resulting 1 of the minuend, the difference is 0 with no borrow. Binary fractions and binary mixed numbers are subtracted in the same way, taking care to place binary points below each other.

1-1-11 Binary complements

As in the decimal system, binary subtraction can be performed as an addition process using the complement method. There are two types of binary complements, the *one's complement* and the *two's complement*.

One's complement. The one's complement of a binary number is found by subtracting each digit of the number from 1 or, more simply, by reversing the digits of the number. For example to find the one's complement of 10101:

Subtraction Method		*Reversing-Digits Method*	
11111		10101	(binary)
− 10101	(pure binary)	↑↑↑↑↑	
01010	(one's complement)	01010	

Although simple, the one's-complement system has certain problems and is not generally used in digital electronic systems. However, the one's complement number must be found as the first step in finding a two's complement number.

Two's complement. The two's complement system is in general use for digital circuits. To convert a pure binary number (which is considered to be positive) to its positive equivalent in two's complement, simply add a zero (or sign bit, as discussed in Sec. 1-1-6) as the next-higher-significant-bit position. For example:

b_7	b_6	b_5	b_4	b_3	b_2	b_1	b_0	
	1	1	1	1	1	1	1	(pure binary 127)
0	1	1	1	1	1	1	1	(+127 in two's complement)
	0	0	0	0	0	0	1	(binary 1)
0	0	0	0	0	0	0	1	(+1 in two's complement)

To convert a pure binary number (considered to be positive) to its negative equivalent in two's complement, invert all the bits (that is, find the one's complement) and then add 1 to the LSD. For example, to convert decimal +37 to negative two's complement:

$$\begin{array}{rl} & \underset{\uparrow\uparrow\uparrow\uparrow\uparrow\uparrow\uparrow\uparrow}{00100101} = +37 \\ \text{invert} & 11011010 \\ \text{add} & \underline{\qquad 1} \\ & 11011011 = -37 \text{ in two's complement} \end{array}$$

Since the positive form of two's complement is the same as pure binary with a positive sign added, the same procedure can be used for conversion. That is, when the negative of a positive two's-complement number is required, the negative is formed by complementing each bit position of the positive representation and then adding 1, as follows:

b_7	b_6	b_5	b_4	b_3	b_2	b_1	b_0	
0	1	1	1	1	1	1	1	(+127 in two's complement)
1	0	0	0	0	0	0	0	(one's complement)
							1	(add 1)
1	0	0	0	0	0	0	1	(−127 in two's complement)
0	0	0	0	0	0	0	0	(0 in two's complement)
1	1	1	1	1	1	1	1	(one's complement)
							1	(add 1)
0	0	0	0	0	0	0	0	(0 is the same in either notation)
0	0	0	0	0	0	0	1	(+1 in two's complement)
1	1	1	1	1	1	1	1	(one's complement)
							1	(add 1)
1	1	1	1	1	1	1	1	(−1 in two's complement)

Note that while +127 is the largest positive two-complement number that can be formed with eight digits, the largest negative two's-complement number is 10000000, or −128. Thus, with two's complement, an 8-bit byte can represent whole numbers between −128 and +127. Bit 7 can be regarded as a sign bit (if b_7 is 0, the number is +; if b_7 is 1, the number is −):

$$\frac{10000000}{-128} \sim \frac{11111111}{-1} \quad \frac{00000000}{0} \quad \frac{00000001}{+1} \sim \frac{01111111}{+127}$$

1-1-12   Relationship between two's complement and signed binary

Since much of the literature on arithmetic operations in digital electronics presents the information in terms of signed binary numbers (Sec. 1-1-6), the difference between two's complement and signed binary is of interest. As discussed in Sec. 1-1-6, signed-binary-number notation also uses the most significant bit as a sign bit (0 for positive, 1 for negative). The remaining bits represent the value as a binary number. For example:

±	64	32	16	8	4	2	1	
b_7	b_6	b_5	b_4	b_3	b_2	b_1	b_0	
1	1	1	1	1	1	1	1	(−127 in signed binary)
1	0	0	0	0	0	0	1	(−1 in signed binary)
0	0	0	0	0	0	0	0	(0 in signed binary)
0	0	0	0	0	0	0	1	(+1 in signed binary)
0	1	1	1	1	1	1	1	(+127 in signed binary)

As 8-bit byte in signed binary represents whole numbers between −127 and +127:

$$\underset{-127}{\underline{11111111}} \sim \underset{-1}{\underline{10000001}} \quad \underset{0}{\underline{00000000}} \quad \underset{+1}{\underline{00000001}} \sim \underset{+127}{\underline{01111111}}$$

Comparing this to the two's-complement representation, the positive numbers are identical and the negative numbers are reversed (−127 in two's complement is −1 in signed binary, and vice versa). In digital circuits, the difference between the two number systems causes no particular problem since numerical information is usually converted (automatically) to the correct format when the circuits are programmed. For example, a typical microcomputer program will provide for the conversion as follows:

1	1	1	1	1	1	1	1	(−127 in signed binary)
1	0	0	0	0	0	0	0	(one's complement except for sign bit)
							1	(add 1)
1	0	0	0	0	0	0	1	(−127 in two's complement)

1-1-13   Binary subtraction using two's complement

Section 1-1-10 describes how binary numbers are subtracted on paper. In most digital electronic systems, binary numbers are subtracted by adding the two's complement of the subtrahend to the minuend, and ignoring the carry bit. This arrangement simplifies circuits of the *arithmetic logic unit* (ALU) found in computers and other digital devices (such as microprocessors). (ALU circuits are discussed in Chapter 3.) The following shows a comparison between conventional binary subtraction (on paper) and two's-complement subtraction (typical of an ALU circuit). Assume that decimal 3 is to be subtracted from decimal 10:

Conventional *(on Paper)*		*Two's Complement* *(in an ALU Circuit)*	
00001010	(decimal 10)	00001010	
− 00000011	− (decimal 3)	+ 11111101	(two's complement
00000111	(decimal 7)	+ 00000111	of decimal 3)

The final carry bit (ninth bit) that results from adding these two binary numbers is not needed. Thus, the bit is not carried, but is ignored and lost. In an 8-bit digital circuit, the circuits are simply not capable of carrying more than 8 bits, so the ninth bit disappears.

1-1-14 Binary multiplication

There are three rules for binary multiplication:

$$0 \times 0 = 0$$
$$0 \times 1 = 0$$
$$1 \times 1 = 1$$

The binary multiplication process is simple. If the digit of the multiplier is 0, the partial product is 0. If the multiplier digit is 1, the partial product is the same as the multiplicand. Then, the partial products are summed as follows:

Binary						Decimal
1 0 1 0 1	(multiplicand)					21
× 1 0	(multiplier)					× 2
0 0 0 0 0	(first partial product)					
1 0 1 0 1	(second partial product)					
1 0 1 0 1 0	(final product)					42

From this it will be seen that multiplication in binary numbers is a form of addition and shifting (the partial product is shifted to the left for each digit in the multiplier; then the products are added). In a digital electronic system, the addition is done by means of *adder circuits,* and shifting is done with *shift registers.* (Operation of adder and shifter circuits is discussed in Chapters 2 and 3.)

Keep in mind that shift operations are used to multiply binary numbers by *powers of 2* (not multiples of 2). A left shift of one position multiplies by 2; a left shift of two bit positions multiplies by 4; three bit positions multiplies by 8; and so on. Similarly, as discussed in Sec. 1-1-15, a right shift of one position divides by 2 (that is, multiplies by one-half); a right shift of two positions divides by 4, and so on.

The same process applies to multiplication of fractions, or mixed numbers.

1-1-15 Binary division

Binary division is performed in much the same way as decimal long division. The process is much simpler, since there are only two rules in binary division:

$$0 \div 1 = 0$$
$$1 \div 1 = 1$$

Generally, division in a digital circuit is a series of subtractions and right shifts performed by subtractor circuits and shift registers (as discussed in Chapter 3). The following example shows the binary division process, as performed on paper:

$$
\begin{array}{cc}
\textit{Binary} & \textit{Decimal} \\[4pt]
\begin{array}{r}
1100 \\
\overline{} \\
1011\,\overline{)10000100} \\
\underline{1011} \\
\overline{1011} \\
\underline{1011}
\end{array}
&
\begin{array}{r}
12 \\
\overline{} \\
11\,\overline{)132} \\
\underline{11} \\
\overline{22} \\
\underline{22}
\end{array}
\end{array}
$$

1-2   OCTAL NUMBER SYSTEM

The octal number system uses eight digits from 0 to 7 in ascending order. Digital computers (particularly older, non-microprocessor-based computers) often use octal numbers in their circuits since octal can provide a shorthand method for bridging the gap between decimal and binary. In octal, when a count is added to the largest digit, 7, the count goes back to the start, 0, and a carry is added to the next position to the left. Thus, decimal 8 appears as octal 10 (pronounced "one, zero," not "ten"). The following list shows the octal equivalents of the 10 digits of the decimal system.

Octal	Decimal
0	0
1	1
2	2
3	3
4	4
5	5
6	6
7	7
10	8
11	9

As in other systems, the digits of an octal number are arranged in positions of ascending powers, from right to left. The following shows the positional values of the octal system.

Digital Position

8^3	8^2	8^1	8^0	8^{-1}	8^{-2}	8^{-3}	powers of 8
512	64	8	1	$\frac{1}{8}$	$\frac{1}{64}$	$\frac{1}{512}$	decimal equivalents

Thus, octal number 1305.1 really means $(1 \times 512) + (3 \times 64) + (0 \times$

8) + (5 × 1) + (1 × 0.125), which is equal to 709.125 in the decimal system.

1-2-1 Conversion from decimal to octal numbers

To convert a decimal number to its octal equivalent, divide the decimal number by 8 and record the remainder, even if it is 0, which becomes the LSD of the octal number. Divide the quotient by 8 again, and record the remainder of this division, which becomes the next-significant octal digit. Continue this process until the quotient becomes 0. The remainder from this division process is the MSD of the octal number. For example, to convert 759_{10} to its octal equivalent:

$759 \div 8 = 94$ with remainder 7 (LSD)
$94 \div 8 = 11$ with remainder 6
$11 \div 8 = 1$ with remainder 3
$1 \div 8 = 0$ with remainder 1 (MSD) $\longrightarrow 1$ $\rightarrow 3$ $\rightarrow 6$ $\rightarrow 7$

Thus, $759_{10} = 1367_8$.

1-2-2 Conversion from octal to decimal numbers

To convert an octal number to its decimal equivalent, multiply each digit by its positional value expressed in decimal equivalents and add the products. For example:

$$123_8 = (1 \times 8^2) + (2 \times 8^1) + (3 \times 8^0)$$
$$= (1 \times 64) + (2 \times 8) + (3 \times 1)$$
$$= 83_{10}$$

1-2-3 Conversion from octal to binary numbers

To convert an octal number to its binary equivalent, convert each digit of the octal number to the binary equivalent *using three binary digits per octal digit*. Then combine these binary groups in proper order. For example, to convert 377_8 to its binary equivalent:

| 3 | 7 | 7 | (octal digits) |
| 011 | 111 | 111 | (binary equivalents) |

Thus, $377_8 = 011111111_2$. As you can see, this grouping of three binary digits will not work for an octal 10 or 11 (decimals 8 and 9, respectively). Also, for three octal digits, it is necessary to use nine binary digits. For these reasons, the octal number system is generally not compatible with the typical 8-bit byte used in microprocesser-based digital equipment.

1-2-4   Conversion from binary to octal numbers

To convert a binary number to its octal equivalent, divide the digits of the binary number into groups of three, starting from the right. If necessary, fill out the last group (at the left) by placing 0s in front. Then convert each binary group to its octal digit equivalent and combine these digits in the proper order. For example, to convert 10110110_2 to its octal equivalent:

added zero⟶ $\boxed{0}10$ 110   110   (binary)

Thus, $10110110_2 = 266_8$.

1-3   HEXADECIMAL NUMBER SYSTEM

One problem with binary numbers is that they are difficult to manipulate, particularly when large values are involved. The length of a large binary number makes it tedious to write and thus more vulnerable to error. This problem is overcome by use of the octal number system, described in Sec. 1-2, where three binary digits are represented by one octal digit. The hexadecimal (or hex) number system goes one step further; *each digit in hex represents four digits in binary.*

Unlike the base 10 of decimal numbers, base 2 of binary numbers, or base 8 of octal numbers, the *hex number system uses a base of 16.* Since the familiar decimal system has only 10 digits, six additional digits are required for the hex system. The first six letters of the English alphabet (A, B, C, D, E, and F) are used. Letter A represents a decimal value of 10, letter B a decimal value of 11, and so on, with letter F representing decimal 15. This is shown in column 4 of Fig. 1-1.

Note that the carry occurs at 16 in hex numbers. Additional digits are required when the count goes beyond 16. For example, a decimal 16 is hex 10 (pronounced "one, zero"), decimal 17 is hex 11 (one, one), and so on. Thus, it is possible to represent a large decimal number with a hex number of fewer digits. For example, $FF_{16} = 255_{10}$ and $FFFF_{16} - 65,535_{10}$.

1-3-1   Conversion from hex to decimal numbers

The simplest way to convert from hex to decimal is to use the table of Fig. 1-1. Note that there are four columns in the table. Column 4 represents the LSD of a four-digit hex number, column 3 is the next-higher-order digit, column 2 is the next-higher digit, and column 1 is the MSD of the four-digit hex number.

To convert from hex to decimal, simply add the decimal equivalents

1		2		3		4	
Hex-Dec		Hex-Dec		Hex-Dec		Hex-Dec	
0	0	0	0	0	0	0	0
1	4,096	1	256	1	16	1	1
2	8,192	2	512	2	32	2	2
3	12,288	3	768	3	48	3	3
4	16,384	4	1,024	4	64	4	4
5	20,480	5	1,280	5	80	5	5
6	24,576	6	1,536	6	96	6	6
7	28,672	7	1,792	7	112	7	7
8	32,768	8	2,048	8	128	8	8
9	36,864	9	2,304	9	144	9	9
A	40,960	A	2,560	A	160	A	10
B	45,056	B	2,816	B	176	B	11
C	49,152	C	3,072	C	192	C	12
D	53,248	D	3,328	D	208	D	13
E	57,344	E	3,584	E	224	E	14
F	61,440	F	3,840	F	240	F	15

Figure 1-1 Basic hexadecimal-decimal conversion table.

for each of the hex digits. Using the number $3C7_{16}$, find the decimal equivalent of hex 3 in column 2, or 768_{10}. Add this to the decimal equivalents of hex C (in column 3, or 192_{10}) and hex 7 (in column 4, or 7_{10}). Thus, $3C7_{16} = 768 + 192 + 7 = 967_{10}$.

1-3-2 Conversion from decimal to hex numbers

The simplest way to convert from decimal to hex is to again use the table of Fig. 1-1. First find the next-lowest decimal number in one of the columns and note the corresponding hex digit. Then move to the nearest column to the right and find a decimal number that, when added to the previous decimal number, will come nearest to the desired decimal number. Note the corresponding hex digit in this column. Repeat the procedure until the decimal numbers, when added, are equal to the desired decimal number. Note all the corresponding hex digits.

For example, to convert from decimal 4011 to hex using Fig. 1-1, note that 3840_{10} in column 2 is the next-lowest decimal number and that F is the corresponding hex digit. Thus, F is the MSD for our hex equivalent number, and the number will have two more digits. Moving to column 3 (the next column to the right), note that 160_{10}, when added to the 3840_{10} (for a total of 4000), will come nearest to the desired 4011. (If 176_{10} in column 3 were chosen, the total is 4016 and thus higher than the desired 4011.) The hex equivalent digit for 160_{10} in column 3 is A. Moving to column 4, note that 11_{10}, when added to the previous $3840 + 160$ (which equals 4000), will equal the desired 4011 exactly and that the hex equivalent digit is B. Thus, $4011_{10} = FAB_{16}$.

1-3-3 Conversion from hex to binary numbers

To convert a hex number to its binary equivalent, convert each digit of the hex number to its binary equivalent *using four binary digits per hex digit*. Then combine these binary groups in proper order.

For example, to convert $3C7_{16}$ to its binary equivalent,

3	C	7	(hex)
0011	1100	0111	(binary)

Thus, $3C7_{16} = 001111000111_2$.

From this, it can be seen that hex is a very convenient shorthand notation which can be substituted for binary. Many present-day digital electronic systems use 8-bit bytes (particularly microprocessor-based devices), and thus only require two hex digits. That is, the 8-bit byte can be divided into two 4-bit nibbles. The left nibble represents the left (MSD) hex digit, and the right nibble represents the right (LSD) hex digit. For example, 115_{10} or 01110011_2 can be written

0111	0011
7	3

When an 8-bit byte is involved, the smallest hex number is 00_{16} (00000000_2) and the largest is FF_{16} (11111111_2). Keep in mind that the digital circuit still reads binary numbers only. Hex is the user's shorthand, used outside the circuit.

1-3-4 Conversion from binary to hex

To convert a binary number to its hex equivalent, divide the digits of the *binary number into groups of four,* starting from the right. If necessary, fill out the last group (at the left) by placing 0s in front. Then convert each binary group to its hex-digit equivalent and combine these digits in proper order. For example, to convert 11011001011_2 to its hex equivalent:

added zero $\longrightarrow$ 0110	1100	1011	(binary)
6	C	B	(hex)

1-3-5 Manipulating hex numbers

Hex addition is a form of counting, as in other number systems. However, since the carry function occurs at 16, the carry must be added when the sum of digits is greater than F (decimal 15). For example, to add hex 7 (decimal 7) and hex A (decimal 10), the sum is hex 11 (decimal 17).

Hex subtraction is similar to that in other number systems except that the borrow occurs at 16. Hex multiplication is similar to decimal multiplication. However, since the base is 16, this must be included when the final

14

product is greater than F (decimal 15). For example, 3 multiplied by 8 is 24 in decimal, but this is represented as 18 in hex. When there is more than one digit in either the multiplier or multiplicand of a hex number, the digits must be handled one at a time, with a shift for each digit. Then the product of the two digits is added. Hex division is accomplished by means of subtraction and shifting, as in other number systems.

From this it can be seen that manipulating hex numbers on paper is a difficult task, especially for those familiar with decimal. Manipulating hex numbers in a digital circuit is equally complex, but fortunately never occurs. In digital circuits, all nonbinary numbers (including hex numbers) are converted to binary before they are added, subtracted, multiplied, divided, and so on. Then the number is converted back from binary as necessary.

For example, assume that a microcomputer is operated with a terminal that has a hex keyboard and a decimal readout. When a hex key is pressed, the number is then manipulated (added, subtracted, etc.) and the result is obtained in binary. This binary result is then converted to decimal for display on the readout.

1-4 BINARY CODES

There are a number of codes based on the binary number system. The simplest form of such coding is where decimal numbers (0 to 9) are converted into binary form using only four binary digits or bits. This *4-bit system* is one of the original codes used in early digital computers, and is still used by some business-oriented systems. With this system, generally known as *binary-coded decimal* or BCD, decimal 1 is represented by 0001, decimal 2 by 0010, and so on.

When the decimal number has more than one digit, 4 binary bits are used for each decimal digit. For example, the decimal number 3738 is represented by 16 binary bits, in groups of 4, as follows:

$$
\begin{array}{cccc}
3 & 7 & 3 & 8 \quad \text{(decimal)} \\
0011 & 0111 & 0011 & 1000 \quad \text{(BCD)}
\end{array}
$$

In a typical 8-bit digital system (such as an 8-bit microcomputer), each byte can be thought of as containing two 4-bit BCD numbers. With this interpretation, each byte can represent numbers in the range from 0 to 99 (decimal). This is shown as follows:

2^3	2^2	2^1	2^0	2^3	2^2	2^1	2^0	
b_7	b_6	b_5	b_4	b_3	b_2	b_1	b_0	
0	0	0	0	0	0	0	0	(decimal 0)
0	0	1	1	1	0	0	0	(decimal 38)
1	0	0	1	1	0	0	1	(decimal 99)

Decimal	Binary	Octal	Hexadecimal	BCD	Reflected gray	2 out of 5	Biquinary 5043210	2421	5421	XS3
0	0000	0	0	0000	0000	00011	0100001	0000	0000	0011–0011
1	0001	1	1	0001	0001	00101	0100010	0001	0001	0011–0100
2	0010	2	2	0010	0011	00110	0100100	0010	0010	0011–0101
3	0011	3	3	0011	0010	01001	0101000	0011	0011	0011–0110
4	0100	4	4	0100	0110	01010	0110000	0100	0100	0011–0111
5	0101	5	5	0101	0111	01100	1000001	1011	1000	0011–1000
6	0110	6	6	0110	0101	10001	1000010	1100	1001	0011–1001
7	0111	7	7	0111	0100	10010	1000100	1101	1010	0011–1010
8	1000	10	8	1000	1100	10100	1001000	1110	1011	0011–1011
9	1001	11	9	1001	1101	11000	1010000	1111	1100	0011–1100
10	1010	12	A	0001–0000	1111			0001–0000	0001–0000	0100–0011
11	1011	13	B	0001–0001	1110			0001–0001	0001–0001	0100–0100
12	1100	14	C	0001–0010	1010			0001–0010	0001–0010	0100–0101
13	1101	15	D	0001–0011	1011			0001–0011	0001–0011	0100–0110
14	1110	16	E	0001–0100	1001			0001–0100	0001–0100	0100–0111
15	1111	17	F	0001–0101	1000			0001–1011	0001–1000	0100–1000

Figure 1-2 Summary of codes used in digital electronics.

There are also many other codes using the binary system, including the 2421, 5421, XS3, reflected gray, 2 out of 5, and biquinary. Such codes are summarized in Fig. 1-2. However, only the BCD is used to any extent. Going further, in present-day digital systems, the trend is to use only hex outside the system, and binary inside the circuits. This is because binary is not compatible with the signal pulses used in digital circuits, and conversion between binary and hex is relatively simple. Also, it is possible to have as many as 256 coded characters with only two hex digits (since each hex digit represents up to 16 possibilities).

1-5 ALPHANUMERIC CODES

While the 4-bit system is adequate to represent any decimal digit from 0 to 9, additional bits are necessary signs to represent letters of the alphabet and special characters (such as dollar signs, percent symbols, etc.) that are often required for digital applications (particularly computers). Most digital equipment manufacturers have settled on the United States American Standard Code for Information Interchange, or USASCII, which is now generally ASCII (pronounced "askey"). Although digital circuits are rarely, if ever, designed to accommodate ASCII directly, there are adapter circuits (decoders) that convert the digital equipment binary code into ASCII (and vice versa) for interchange of information with the world outside the digital system. (Decoders are discussed in Chapter 2.)

ASCII is an 8-bit code and is thus ideally suited for hex representation. Also, since hex–binary conversion is relatively simple, both on paper and in the digital circuits, ASCII can be adapted to any digital system (and to microprocessor-based systems in particular).

Figure 1-3 shows the conversion between ASCII and hex. To convert from ASCII to hex, select the desired letter, symbol, or number, then move up vertically to find the hex MSD. Then move horizontally to the left and find the hex LSD. For example, to find the hex code for the letter I, note that I appears in the "4" column of the hex MSD and in the "9" column of the hex LSD. Thus, hex 49 equals the letter I in ASCII. Going further, the hex 49 can be converted to 0100   1001 in binary, as is generally done inside digital circuits. The process can be reversed to convert from binary to ASCII. For example, binary 0010   0100 is 24 in hex, and $ in ASCII. As an exercise, find the ASCII letters for binary 0100-1100   0100-0101   0100-1110   0100-1011.

Keep in mind that all digital electronic equipment does not use ASCII. Thus, although any 8-bit byte in binary can be converted to hex for convenience, the result does not necessarily mean anything in ASCII. Also, when ASCII is used, the ASCII characters are converted to binary for use by the digital circuits, and then converted back to ASCII for use by the outer world (such as a computer terminal). Also, as shown in Fig. 1-3, there are some variations in ASCII-coded devices but not in the basic format. For example, a Model 33 teletype prints codes in MSD columns 6 and 7 as if they were column 4 and 5 codes. Such factors must be considered when interfacing digital equipment. Interfacing is discussed further in Chapter 3.

1-6 RELATIONSHIP BETWEEN BINARY NUMBER SYSTEMS AND DIGITAL PULSES

Thus far, we have discussed how binary numbers and words are used on paper. Digital equipment operates with electrical signals (generally pulses) arranged in binary form. Let us take the example of a microprocessor, which is the heart of most present-day microcomputers, as discussed in Chapter 4. A microprocessor performs its functions (program counting, addition, subtraction, etc.) in response to *instructions*. Usually, these instructions come from a *memory* within the system, but they can also come from the outside world via a terminal. A typical microprocessor can perform from 70 to 100 functions (or possibly many more), with each function being determined by a specific instruction.

The instructions are applied to the microprocessor as electrical pulses,

Most-significant hex digit

		0	1	2	3	4	5	6	7
	0	NUL	DLE	SP	0	@	P	\	P
	1	SOH	DC1	!	1	A	Q	a	q
	2	STX	DC2	''	2	B	R	b	r
	3	ETX	DC3	#	3	C	S	c	s
	4	EOT	DC4	$	4	D	T	d	t
	5	ENQ	NAK	%	5	E	U	e	u
Least-significant hex digit	6	ACK	SYN	&	6	F	V	f	v
	7	BEL	ETB	/	7	G	W	g	w
	8	BS	CAN	(	8	H	X	h	x
	9	HT	EM	)	9	I	Y	i	y
	A	LF	SUB	*	:	J	Z	j	z
	B	VT	ESC	+	;	K	[	k	{
	C	FF	FS	,	<	L	\	l	¦
	D	CR	GS	−	=	M	]	m	}
	E	SO	RS	.	>	N	↑	n	~
	F	SI	US	/	?	O	←	o	DEL

Notes:

(1) Parity bit in most-significant hex digit not included.

(2) Characters in columns 0 and 1 (as well as SP and DEL) are non printing.

(3) Model 33 teletype prints codes in columns 6 and 7 as if they were column 4 and 5 codes.

Figure 1-3 ASCII–hexadecimal conversion table.

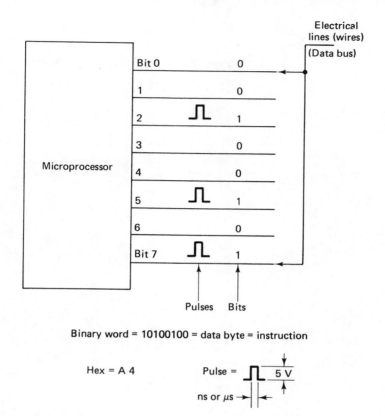

Figure 1-4 Relationship between binary numbers and electrical signals.

arranged to form a binary word. Each pulse is applied on a separate electrical line (or wire) as shown in Fig. 1-4, where a basic microprocessor has eight lines to accommodate an 8-bit binary word or *data byte*. In this system, which is typical for the great majority of microprocessors, the pulses are +5 V in amplitude and a few ns or μs in duration. The presence of a pulse indicates a binary 1; the absence of a pulse (or zero volts) indicates a binary 0. Thus, in Fig. 1-4, the microprocessor is receiving a binary 10100100, which can be converted to hex A 4 for convenience.

As is discussed in Chapter 4, this word may be an instruction to perform addition for one particular microprocessor. For another microprocessor, the same words may mean "perform subtraction." For still another microprocessor, the word may be meaningless. However, before we get further into microprocessors, let us consider some basic digital logic, as discussed in Chapter 2.

CHAPTER

2

Basic
Digital Logic

Digital electronic equipment is based on the use of digital *logic circuits*. In turn, logic circuits are based on the use of *logic equations*. To fully understand digital electronics, you must therefore understand logic equations, and logic circuits that are implemented with *logic gates*. As a practical matter, to understand digital electronics in its present-day form, you must also understand the use of logic circuits found in IC modules, since most digital equipment is now made up of ICs. Chapter 3 discusses typical logic circuits, including digital ICs. In this chapter we concentrate on digital logic basics.

No matter what logic elements are used, many problems in digital design can be solved by working out the solution in equation form first, and then making circuits (or interconnecting digital IC elements) to match the equations. As an example, if a circuit is to be *minimized* (reduced to the least number of logic elements), the digital equipment is reduced to the simplest form "on paper" and the simplified equation is converted to a circuit. Note that logic equations may also be known as logical algebra, Boolean algebra (after the English mathematician George Boole), computer algebra, or computer logic.

If you are involved in any phase of digital design or experimentation, you should be able to manipulate logic equations (usually to simplify them) and then to convert the equations into practical circuits. This task is greatly simplified if you can write an equation for a given circuit as a first step toward improving the circuit. Even when simplification is not involved, it is easier to write an equation than to wire a circuit. A good example of this is a logic module that has six inputs and one output, in which the output must be present when three (and only three) of the inputs are present. With such a problem, the equation is written to express the relationship of the six inputs to the output. Then the equation is converted to a corresponding circuit. Another classic example is where it is necessary to convert a digital circuit from positive logic to negative logic in order to accommodate an external circuit function.

In addition to a practical working knowledge of logical algebra, you should be familiar with the *symbols* used in logic circuits, the binary and other digital number systems (discussed in Chapter 1), the basic *logic circuit forms* (including *combinational logic* and *sequential logic*), *mapping* of logic circuits, and certain other problems commonal to all digital logic design. These subjects are discussed in summary form throughout this chapter.

2-1 DEFINITIONS, STATES, AND QUANTITIES OF LOGICAL ALGEBRA

Logical algebra is a method of symbolically expressing the relationship between logic variables. This is in contrast to conventional algebra, which is the symbolic expression for relationships of number variables. Logical algebra differs from conventional algebra in two respects:

1. The symbols used in logical algebra (usually letters) do not represent numerical values.
2. Arithmetic operations are not performed in logical algebra.

Logical algebra is ideally suited to any system of intelligence based on *two opposite states* such as "on" or "off." Thus, logical algebra is specifically suited to express the opening and closing of electrical switches, the presence or absence of electrical pulses, or the polarity or amplitude relationship of pulses. Logical algebra is also quite compatible with the binary number system (and the various binary-based codes) discussed in Chapter 1.

The following is a summary of the states and quantities available in logical algebra.

1. *Logical algebra has only two discrete states.* Any pair of conditions, different from each other, can be chosen. In digital electronics, the

states are described as "true" or "false," or possibly "up" or "down" (or "high" or "low," referring to the presence or polarity of pulses. You can assign any dissimilar value or state to represent "true" and any other value to represent "false."

2. *Logical quantities are single-valued.* No quantity in logical algebra can be simultaneously both "true" and "false." If the quantity is "false," then the opposite quantity is "true."

3. *A logical quantity may be either constant or variable.* Any quantity that is "true" is equal to any other quantity that is "true." If constant, a logical quantity must remain "true" or "false." If variable, the quantity may switch between the "true" and "false" states from time to time, but only between the *two extremes.*

4. *There are many ways to represent logical algebra in digital electronics.* To help explain operation of digital circuits, some textbooks use the presence or absence of light ("lamp on" or "lamp off") or the opening and closing of switches to represent the true and false states of logical algebra. However, in practical digital electronic circuits, the states are usually represented by the presence or absence of electrical pulses. (In some systems, the amplitude or polarity of pulses is used to identify the logical algebra true and false states.) In one of the most common systems, the presence of a pulse represents the "true" state or quantity, and corresponds to the "1" state in the binary number system (Chapter 1). With this system, the absence of a pulse represents "false" (or 0 in binary). However, another system might use the exact opposite representation ("false" or 0 by the presence of a pulse; "true" or 1 by the absence of a pulse).

2-2 NOTATION IN LOGICAL ALGEBRA

The condition or state of a logic variable in logical algebra are generally represented by letters of the alphabet. For example, assume that the letter A represents the condition of a logic variable. In that case, the values A can assume will be only true or false, since these are the only values that any logic variable may represent. (This is quite different from conventional algebra, where the letter A can represent any value from minus to plus infinity.)

The inverse, or *complement,* of the letter A is represented by the symbol $\overline{A}$ (A-bar or A-overbar), A' (A-prime), or A* (A-asterisk or A-star). The prime and asterisk signs are the most popular notations in text for the complement condition, since they are available on most typewriters. However, on digital electronic diagrams, the overbar is most popular.

No matter what system of notation is used to show the complement, or inverse, condition, A and $\overline{A}$ cannot have the same value at the same time. Thus, if A is false at any given time, then $\overline{A}$ must be true, and vice versa.

If two or more logic variables are present at the same time, one might be represented by A, another by B, and so on. If at this time B happens to be true and A happens to be true, then A = B, since a true always equals a true. Simultaneously, the complement of B (or $\overline{B}$) and the complement of A (or $\overline{A}$) are equal to each other, and are false. A moment later, the variable represented by A may change state, and A then becomes false and $\overline{A}$ (the complement) becomes true.

It is important to remember that A, $\overline{A}$, B, and so on, are simply symbols used to represent logic variables. Also, at any time any symbol may be true, and a moment later the same symbol may be false.

In older logic diagrams, the symbol or letter T is often used to mean true, with F used to indicate false. In other cases, F is used to indicate function or output. In most present-day logic diagrams, 1 represents true and 0 represents false. The 1 and 0 are in common use since they are similar to binary numbers and are easy to write. Note that 1 in logical notation is not necessarily equal to the number 1, and 0 is not necessarily equal to the number 0. The symbols 1 and 0 only denote true and false states in logical variables.

For example, 1 means *yes, assertion, up, high, enable,* or *true.* Consequently, 0 means *no, negation, down, low, disable* (or *inhibit*), or *false.* Use of the words *true* and *false* (or *up* or *down*) does not imply that one state is more important than the other, or that one state is necessarily the "normal" state. The states are conditions, and both states are equally significant and are used equally in a two-state system.

Assume that a digital diagram shows a circuit using − 5 V to represent true, or binary 1, and 0 V to represent false, or binary 0. Under these conditions, the 4-bit binary number 0111 (decimal 7) present at the output of a digital circuit could be represented electronically on four different lines, as shown in Fig. 2-1.

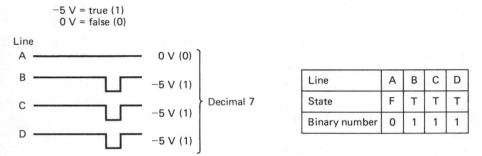

Figure 2-1 Relationship between logical algebra notation, practical circuit voltage levels, and binary numbers.

The basic operations in logical algebra (and the corresponding digital electronic circuits) are AND, OR, and NOT (or invert). In addition, there are five common operations (or digital electronic functions) in logical algebra that are produced by combining the functions of these three basic operations. The common operations are NOR, NAND, AND/OR, EXCLUSIVE OR, and EXCLUSIVE NOR.

As in conventional algebra, certain symbols (or *connectives*) are used to indicate the type of operation that is to be performed in logical algebra. Unfortunately, the use of connective symbols is not standard throughout the industry. However, the following symbols are generally accepted and understood.

The AND operation is represented by a dot between the variables (A·B). In most cases, the dot is omitted but is implied. Thus, ABC is read (A and B and C).

The OR operation is represented by a plus sign between the variables (A + B). This reads: A or B.

The NOT operation is represented by a long bar over the logical quantity to be complemented or inverted. This is similar to the inverse or complement notation. Thus, when A is subjected to the NOT operation, it becomes $\overline{A}$. However, when the complement is subjected to NOT, the complement is removed. Thus, when A is subjected to NOT, it becomes $\overline{\overline{A}}$. Similarly, $\overline{\overline{A}}$ = A.

If the term subjected to the NOT operation contains an OR or an AND operation, in addition to the variables, the OR or AND operation is also complemented (or inverted). That is, if the variables are connected by an AND, the AND is changed to an OR, in addition to changing the variables. For example:

$$\overline{A + B} = AB \qquad \overline{AB} = A + B$$
$$\overline{\overline{A} + B} = A\overline{B} \qquad \overline{A\overline{B}} = A + \overline{B}$$

Thus, the complement of the AND operation is an OR, and the complement of OR is an AND.

The equal sign (=) found in conventional algebra is carried over into logical algebra, and with the same meaning (equals, or, is the result of). However, the equal sign (=) with two bars is a *conditional equal* in logical algebra. That is, the conditions must be stated for the equations to be correct. For example, A = B, where A is true and B is true. The equal sign (≡) with three bars is an unconditional equality. For example, A ≡ B means that A is the same as B, under all conditions. Figure 2-2 shows typical symbols, equivalent symbols, and forms of notation used in digital logic. Figure 2-3

Symbol or notation	Description
· ∩ ∧ X	AND
+ ∪ ∨	OR
Ι ⊼	NAND
↓ ⊽	NOR
⊕	EXCLUSIVE OR
$\overline{\oplus}$	EXCLUSIVE NOR
$\overline{X}$ ~ X X− X′	NOT X
=	Equivalence
A ∪ B	Union of A and B
A ∩ B	Intersection of A and B
A ⊂ B	A belongs to B
A → B	A implies B
A ⊃ B	A implies B
1	True
0	False
A = B	A = B, conditionally
A ≡ B	A is identical to B, unconditionally
A ⊕ B	A EXCLUSIVE OR B
$\overline{A}$	NOT A
A + B	A OR B

Figure 2-2 Typical symbols, equivalent symbols, and forms of notation used in digital logic and Boolean algebra.

shows equations (grouped according to function where practical) commonly used in logical algebra.

2-4 TRUTH TABLES

The truth table is one of the most useful tools for analyzing problems in logical algebra, and to show the functions of the corresponding digital electronic circuits. As shown in Fig. 2-4, the truth table consists of one vertical column for each of the logic variables involved in a given problem. The horizontal lines or rows of the truth table are filled with all possible true–false combinations the variables can assume with respect to each other.

For example, two variables can assume four different combinations at any of four different times: that is, both true at the same time; both false at the same time; and one true and one false. There are no other possible com-

T = true   F = false

$AT = A$
$AF = F$
$AA = A$
$A\overline{A} = F$
$\overline{\overline{A}} = A$ Basic laws
$A + T = T$
$A + F = A$
$A + A = A$
$A + \overline{A} = T$

$A + B = B + A$ Communicative laws
$AB = BA$

$A(B + C) = (AB) + (AC)$ Distributive laws
$A + (BC) = (A + B)(A + C)$

$(A + B) + C = A + (B + C)$ Associative laws
$(AB)C = A(BC)$

$A + (AB) = A$
$A(A + B) = A$ Simplifying theorems
$A + (\overline{A}B) = A + B$
$AB + A\overline{B} = A$

$\overline{A + B + C} = \overline{ABC}$ De Morgan's theorems
$\overline{ABC} = \overline{A} + \overline{B} + \overline{C}$

Figure 2-3 Equations commonly used in logical algebra.

Variables

0–1 table

A	B	Output
0	0	0
0	1	0
1	0	0
1	1	1

0 V = false
+5 V = true

Voltage-level table

A	B	Output
0 V	0 V	0 V
0 V	+5 V	0 V
+5 V	0 V	0 V
+ 5V	+5 V	+5 V

Variables

A	B	AND	NAND	OR	NOR	EXCLUSIVE OR	EXCLUSIVE NOR
0	0	0	1	0	1	0	1
0	1	0	1	1	0	1	0
1	0	0	1	1	0	1	0
1	1	1	0	1	0	0	1

Figure 2-4 Typical two-variable truth table and a summary of truth tables for logic gates.

binations. The additional column (or columns) contain the output or results produced by the combinations of variables. For example, the output column (also known as the "function" column on some digital logic diagrams) can contain the result, or "output," when the variables are ANDed together, ORed together, or any particular output that must occur for each combination stated in the logic problem.

Truth tables can be made up using any combination of letters, numbers, or symbols. Most digital diagrams use 1s and 0s for truth tables. Figure 2-4 shows the truth tables for a positive-logic AND gate using 1s and 0s, as well as the corresponding voltage levels. In Fig. 2-4, 1 is true and is represented by +5 V; 0 is false and is represented by 0 V. Figure 2-4 also summarizes the truth tables for the six basic digital logic gates AND, OR, NAND, NOR, EXCLUSIVE OR, and EXCLUSIVE NOR. Each of these digital gates is discussed in the following paragraphs.

2-5 DIGITAL LOGIC PULSES AND VOLTAGE LEVELS

Although any system of letters and numbers can be used to represent true and false in digital diagrams, *voltage levels and pulses* are used in practical digital circuits. Because digital circuits can work equally well with positive or negative logic, it is necessary to define if the logic is positive or negative. In some (but not all) digital diagrams, a plus or minus sign may be used within a logic symbol to define the *true state* for that element. The sign should be used for all digital logic elements in which true and false levels are meaningful. As an alternative, the logic diagram can state in a note that all logic is positive true or that it is all negative true. Unfortunately, not all digital diagrams contain this notation.

When used, the plus sign within a logic symbol means that the *relatively positive level* of the two logic voltages at which the digital circuit operates is said to be true. This is defined as positive logic. Note that the true voltage level does not have to be absolutely positive (that is, above ground or above the 0-V reference). The two voltage levels at which a logic circuit operates could be −1.7 V and −0.9 V (which is the case for ECL logic described in Chapter 3). A plus sign within the logic symbol (or a note specifying positive true) indicates that the −0.9 V is true and that the −1.7 V is false, since the −0.9 V is closer to positive than −1.7 V.

A minus sign within the symbol (or a note specifying negative true) indicates that the −1.7-V level is true (since it's more negative) and that the −0.9-V level is therefore false. This is defined as negative logic. Note that the voltage levels shown in Fig. 2-1 use negative logic. That is, the true state is at −5 V, whereas the false state is at 0 V.

Note that the voltage shown in Fig. 2-1 are actually electrical pulses

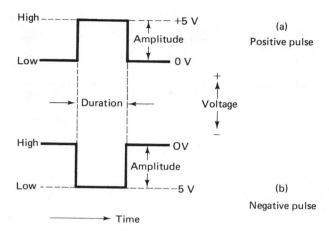

Figure 2-5 Simplified digital pulse definitions.

rather than constant voltage levels. It is important that you understand the relationship of positive and negative logic to pulses, as well as some basic pulse terms. The pulse of Fig. 2-5(a) is a positive pulse since it goes from a lower level to a higher level, and back to a lower level. The pulse of Fig. 2-5(b) is a negative pulse because it goes from a higher level to a lower level, and back to a higher level. Lower and higher levels represent *voltage levels*. Lower represents the lower voltage level or *down* voltage. Higher represents the higher voltage or *up* voltage level. The difference between the lower and higher levels represents the *pulse amplitude* (which is typically 5 V).

With *positive logic,* the upper level of the pulses in Fig. 2-5 is the true condition, with the lower level the false condition.

With *negative logic,* the lower level of the pulses is the true condition, with the upper level the false condition.

With either system, the length of time that a positive pulse stays up or a negative pulse stays down is called the *pulse duration.*

2-6 DIGITAL FUNCTIONS AND OPERATIONS

This section describes the six basic digital functions or operations.

2-6-1 The AND operation

The AND operation is true when all of the ANDed quantities are true, and is false when one or more of the ANDed quantities is false. In the case of a digital AND gate, the output is true, or 1, when all of the inputs are 1, and is 0 when one or more of the inputs is 0.

Figure 2-6 shows the truth table, symbol, and circuit diagrams of a

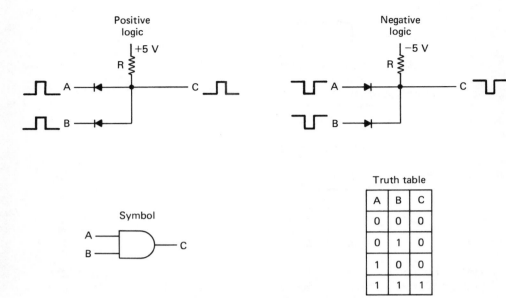

Figure 2-6   AND gate circuit diagrams, symbol, and truth tables.

basic digital AND gate. The diagrams show both positive and negative logic. The circuits shown are often referred to as "diode logic," since only switching diodes (and no transistors) are used. As discussed in Chapter 3, in practical digital circuits, gates are usually found in IC form. As is typical for ICs, the internal circuits are not shown on the diagrams. The symbol is used instead. The internal gate circuits are shown here to help you understand operation of the gates and to show the relationship of positive and negative logic.

With positive logic, when both inputs A and B are 1 (as indicated by the corresponding positive pulses), then and only then the output is 1. (Both diodes are reverse-biased, current flow through resistor R is halted, and the output line rises to some voltage near that of the source of +5 V.) If any one or both inputs are 0, then the AND function is not satisfied, and the output goes to 0. (One or more of the diodes are forward-biased, current flows through resistor R, and the output line drops.) Resistor R is often referred to as the *pull-up resistor* in digital literature.

In practical digital circuits, AND gates may appear in multiple-input form (usually about six inputs is maximum). Also, many thousands of gates may be included in a single IC. However, no matter how many inputs are involved, *all inputs to an individual AND gate* must be 1 (true) for the output to be 1.

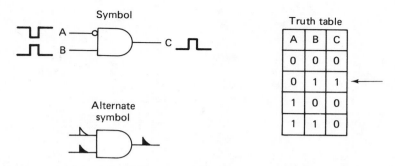

Figure 2-7 Symbol and truth table of AND gate with mixed inputs.

Although all inputs must be true for an AND gate to produce a true output, the inputs need not necessarily be of the same polarity. That is, the internal gate circuit can be arranged to produce a 1 output when mixed 1 and 0 inputs are applied. This is shown in Fig. 2-7. An inverted input is represented on the symbol by a dot (or small circle) on that input which is inverted from the *normal logic* on the rest of the diagram.

As an example, with positive logic, an inversion dot on an input indicates that a 0 input is required to produce a true condition. If it is assumed that the symbol of Fig. 2-7 uses positive logic, a 0 input is required at input A and a 1 at input B to produce a 1 output (as shown by the truth table).

Note that an alternate symbol is shown on Fig. 2-7. The inputs and output of the alternate symbol (which is part of military standard MIL-STD-806C) are provided with arrows or flags. A solid arrow is the equivalent of no inversion dot, and an open arrow is the same as an inversion dot. However, since this alternate system is not in common use (except in the digital diagrams of military manuals using MIL-STD-806C), you can promptly forget the system. In general, MIL-STD-806 has been replaced by ANSI Y32.14/IEEE, even for military use. However, do not be surprised at any symbols that you may find in digital literature.

2-6-2 The OR operation

The OR operation is true when one or more of the quantities is true, and is false only when all of the quantities are false. This is shown in the truth table of Fig. 2-8. In practical digital circuits, OR gates may appear is multiple-input form, and many thousands of gates may be included in a single IC. However, no matter how many inputs are involved, *any one true input to an OR gate will produce a true output.*

As with AND gates, the internal circuits of an OR gate can be designed so that mixed input will produce a true output. For example, if it is assumed

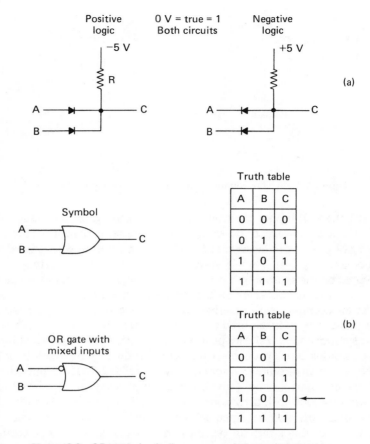

Figure 2-8   OR gate circuit diagrams, symbols, and truth tables.

that the symbol of Fig. 2-8(b) uses positive logic, a 0 input at A or a 1 at B will produce a 1 output.

2-6-3   The AND-OR operation

The AND-OR operation is not basic in the same sense as the AND and OR operations. However, the AND-OR operation is used so frequently in digital circuits (particularly in a version where the OR output is inverted) that the function can be treated as a basic operation.

Figure 2-9 shows the truth table and logic symbol for the AND-OR operation. Note that the AND-OR operation is made up of two AND gates at the input and one OR gate at the output.

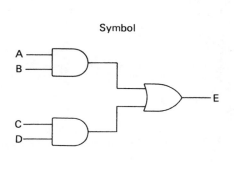

Truth table

A	B	C	D	E
0	0	0	0	0
0	1	0	0	0
1	0	0	0	0
1	1	0	0	1
0	0	0	1	0
0	0	1	0	0
0	0	1	1	1

Figure 2-9 AND-OR operation symbol and truth table.

2-6-4   The NOT operation

The NOT function is often referred to as negation, or a complementing function. The negative of a quantity may also be called the inverse, converse, or opposite. The NOT circuit is required to produce the NAND and NOR gate functions (described in Secs. 2-6-5 and 2-6-6).

The output of a NOT circuit is the inverse of the input. For example, if the input is a positive-going pulse (representing true or 1), the output will be a negative-going pulse (representing false or 0). The NOT function is often combined with amplification.

The basic NOT circuits are shown in Fig. 2-10. Note that these circuits are essentially single-stage, common-emitter transistor amplifiers. In a negative-logic system, a PNP transistor is used, since a negative voltage at the base causes the transistor to conduct, creating a ground (0-V) output that is the opposite of the negative input voltage. For example, assume that the logic is negative, as shown in Fig. 2-10(b), that true is represented by − 5 V, and that false is 0 V. If the input is true, the − 5 V at the base causes the transistor to conduct. The output (collector) then rises to 0 V (or near 0 V), producing a false output. If the input is false, the 0 V at the base does not turn the transistor on, and the output remains at some voltage near the source of − 5 V (or true). An NPN transistor is used with positive logic to produce the corresponding results.

The transistor circuit may also provide some amplification. The output of the diode is always less than the input, so diode gates do not have the capability of driving many other diode circuits. The transistor NOT circuit can provide pulse-amplitude restoration, and driving capability, in addition to inversion of the pulses.

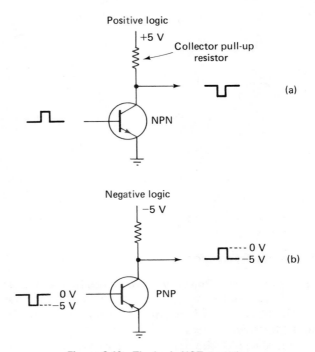

Figure 2-10   The basic NOT operation.

2-6-5   The NAND operation

    The NAND operation is a combination of the NOT and AND operations. The term NAND is a contraction of NOT AND. The NAND output is true when one or more of the inputs is false, and is false only when all of the inputs are true. This is shown in the truth table of Fig. 2-11. A typical NAND gate will include an AND circuit and a NOT amplifier. The AND circuit performs the usual AND function, producing a true output when both inputs

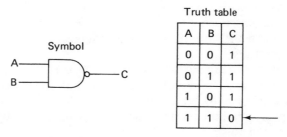

Figure 2-11   NAND gate symbol and truth table.

are true. The NOT amplifier inverts this into a false. Thus, two true inputs produce a false output. If either (or both) inputs is false, the AND circuit produces a false output which is inverted to a true by the NOT circuit. The NAND gate is generally preferred to the AND gate for most logic design applications because of the amplification factor provided by the NOT circuit.

2-6-6 The NOR operation

The NOR operation is a combination of the NOT and OR operations. The term NOR is a contraction of NOT OR. The NOR output is false when one or more of the inputs is true, and is true only when none of the inputs is true. This is shown in the truth table of Fig. 2-12. A typical NOR gate will include an OR circuit and a NOT amplifier. The OR circuit performs the usual OR function, producing a true output when either or both inputs are true. The NOT circuit inverts this true into a false. Thus, one or both true inputs produce a false output. If both inputs are false, the OR gate produces a false output, which is inverted to a true output by the NOT circuit. The NOR gate is generally preferred to the OR gate for most logic design applications because of the amplification factor provided by the NOT circuit.

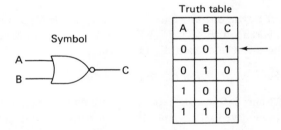

Figure 2-12 NOR gate symbol and truth table.

2-6-7 The EXCLUSIVE OR, EXCLUSIVE NOR, and COINCIDENCE operations

An EXCLUSIVE OR gate is a special type of OR gate. This gate has only two inputs, and one output. The output is true if one, but not both, of the inputs is true. The converse statement is equally accurate: the output is false if the inputs are both true or both false. This is shown by the truth table of Fig. 2-13(a). EXCLUSIVE OR gate is independent of polarity, and is not generally regarded as being either positive-true or negative-true.

An EXCLUSIVE NOR gate operates in the same way as the EXCLUSIVE OR, except that the output is inverted. This is shown in the truth table of Fig. 2-13(b). With EXCLUSIVE NOR, two false inputs, or two true

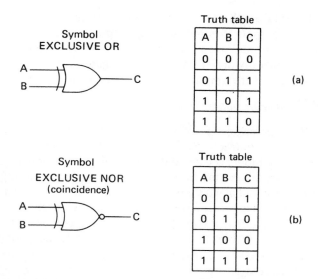

Figure 2-13 EXCLUSIVE OR and EXCLUSIVE NOR symbols and truth tables.

inputs, produce a true output. One false and one true input produce a false output. For this reason, EXCLUSIVE NOR is sometimes called the COINCIDENCE operation in certain digital literature.

2-6-8 Equivalent digital operations

Positive logic defines the 1 or true state as the most positive voltage level, whereas negative logic defines the most negative voltage level as the 1 or true state. Because of the difference in definition of states, it is possible for some digital elements to have two equivalent outputs, depending on definition. Figure 2-14 shows a comparison of several common digital functions or gates. As an example, a positive-logic AND gate produces the same outputs as a negative-logic OR gate, and vice versa. The same relationships exist between NAND and NOR, as well as EXCLUSIVE OR and EXCLUSIVE NOR.

2-7 DIGITAL CIRCUIT SYMBOLS

Although there are many variations of digital elements and their circuit symbols, there are only four basic classes or groups: gates, amplifiers (or inverters), switching elements, and delay elements. In present-day digital devices, most of the elements are contained within ICs, and the symbol does not appear on the diagram. However, since individual digital elements are

Inputs		Positive logic					
A	B	AND	OR	NAND	NOR	EXCLUSIVE OR	EXCLUSIVE NOR
0	0	0	0	1	1	0	1
0	1	0	1	1	0	1	0
1	0	0	1	1	0	1	0
1	1	1	1	0	0	0	1
A	B	OR	AND	NOR	NAND	EXCLUSIVE NOR	EXCLUSIVE OR
Inputs							
		Negative logic					

Figure 2-14 Comparison of positive and negative logic functions.

used in many applications, it is essential that you recognize and understand the corresponding symbols as they appear on diagrams. This section describes the symbols that are unique to digital circuit diagrams. The symbols for other electronic parts (transistors, capacitors, resistors, etc.) are the same as the symbols used on the schematic diagrams of conventional, non-digital, electronic devices.

2-7-1 Gate symbols

A gate is a circuit that produces an output on condition of certain rules governing input combinations. As shown in Fig. 2-15, the basic gate symbol has input lines connecting to the flat side of the symbol, and output lines

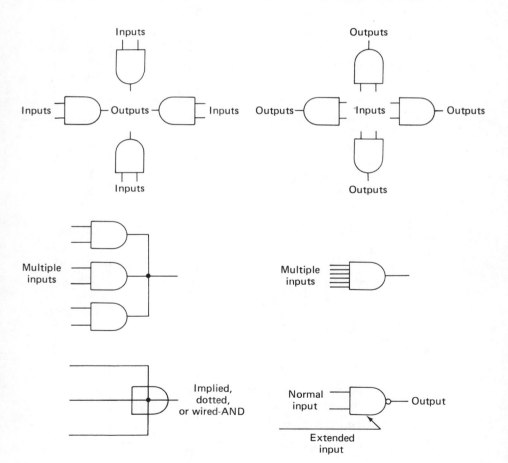

Figure 2-15 Basic gate symbols.

connecting to the curved or pointed sides. Since inputs and outputs are thus easily identifiable, the symbol can be shown facing left or right (or facing up or down), as necessary.

There may be two inputs to a gate. In some cases, multiple inputs (more than two) can be accomplished by increasing the internal circuit's component of the gate (such as adding an extra diode for each input). In other cases, it is necessary to connect the output of two (or more) gates together, in parallel. For example, the output of three two-input AND gates can be connected in parallel, resulting in a six-input AND gate. This is sometimes known as a *wired, dotted,* or *implied* AND function. Digital diagrams often use the symbol shown in Fig. 2-15 when the term WIRED-AND is involved.

It is also possible for a gate to have an input other than the normal input. This is often referred to as an "extended input." For example, a NAND gate can be made up of a diode AND gate, followed by a transistor amplifier (which also inverts). An input can be connected directly to the transistor base, thus bypassing the diode AND function. Generally, a signal at an extended input produces an output regardless of the condition (true or false) at all of the regular inputs.

2-7-2   Amplifier symbols

When amplifiers are used in digital circuits, the driving or input signals are normally pulsed. Consequently, the output of the amplifier is an amplified form of the input pulse. As shown in Fig. 2-16, the amplifier symbol is a triangle with the input applied to the center of one side and the output taken from the opposite point of the triangle. Like gates, the amplifier may be shown in any of the four positions.

When an amplifier is used as a separate element in digital circuits, it is assumed that the output is essentially the same as the input, but in amplified form. That is, a true input produces a true output, and vice versa. When inversion occurs, an inversion dot (or possibly an inverted pulse symbol) is placed at the output. Usually, the element is then termed an *inverter* rather than an amplifier, even though amplification may occur. Also, an amplifier (with or without inversion) can be called a *buffer* when it is used between the two logic elements or circuits.

If a plus or minus sign is used in the symbol, this usually indicates the *input polarity* required to turn the amplifier on.

One amplifier or inverter symbol usually represents many amplification stages. Digital symbols, by themselves, do not necessarily imply a specific number of components, but rather relate to the overall effect. Similarly, amplifiers in digital circuits can have more than one input and output. For example, as shown in Fig. 2-16, a *phase splitter* has one input and two outputs. One of the outputs is in phase with the input, whereas the other

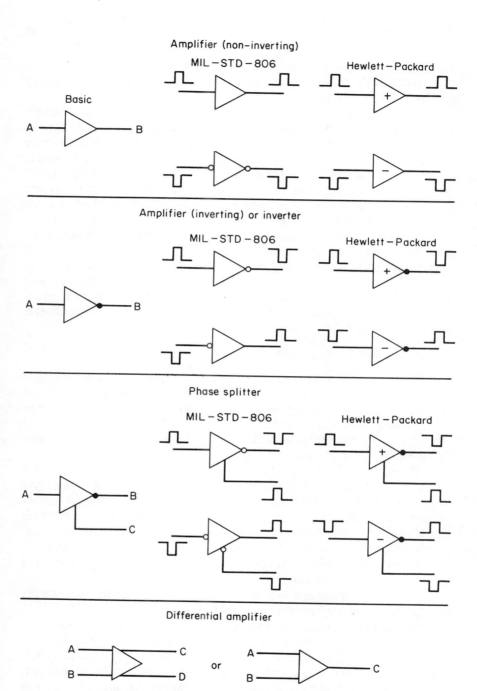

Figure 2-16 Amplifier, inverter, and phase-splitter symbols.

output is out of phase with the input. A similar case exists with *differential amplifiers,* which have two inputs and two outputs (although some differential amplifiers have dual inputs and single outputs).

2-7-3   Switching element symbols

Switching elements used in digital circuits are some form of multivibrator: bistable (flip-flop, latch, Schmitt trigger), monostable (one-shot), and astable (free-running). According to the type of circuit, inputs cause the state of the circuit to switch, reversing the output. For example, what is true will switch to false, and vice versa.

The flip-flop or latch is the most common digital switching element. A flip-flop, or FF, is bistable; that is, it takes an external signal to set the FF, and another signal to reset the FF. The FF will remain in a given state until switched to the opposite state by the appropriate external signal.

The basic digital switching element and FF symbols are shown in Fig. 2-17. The various types of FFs and switching elements used in present-day digital circuits are discussed further in other sections of this chapter, and in other chapters, as applicable.

2-7-4   Delay element symbols

A delay element provides a finite time between input and output signals. The delay symbols, with examples of actual delay time, are shown in Fig. 2-18. Many types of delay elements are used in digital circuits. Two frequently used delays are the *tapped delay* and delays effective only on the leading or trailing edges of pulses. Such delay elements, together with the theoretical waveforms, are shown in Fig. 2-18.

2-7-5   Modifying and identifying digital circuit symbols

Basic digital circuit symbols are often modified to express circuit conditions. Although designers may have their own set of modifiers, together with those of MIL-STD-806, the following modifiers are in general use for digital circuit diagrams.

Truth polarity. As previously discussed, positive $(+)$ or negative $(-)$ indicators may be placed inside a symbol to designate whether the true state for that circuit is positive or negative, *relative to the false state.* This is frequently done with gates and switching elements, as shown in Fig. 2-19. When all symbols on a particular digital diagram have the same polarity, a note to the effect that all logic is positive-true or all is negative-true may be used instead of having individual polarity signs in each symbol. Polarity signs used in amplifier symbols do not usually have any direct logic significance. Rather,

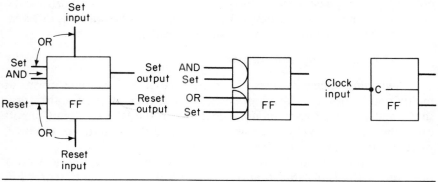

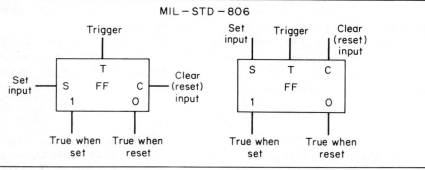

MIL – STD – 806

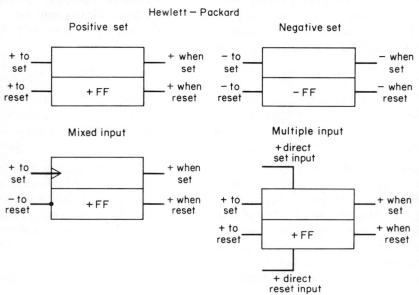

Hewlett – Packard

Figure 2-17 Basic FF symbols.

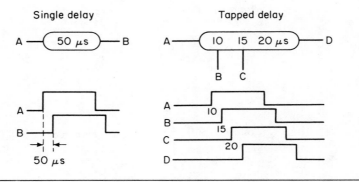

Single delay

Tapped delay

MIL – STD – 806

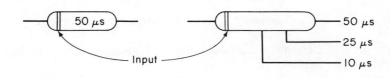

Hewlett – Packard

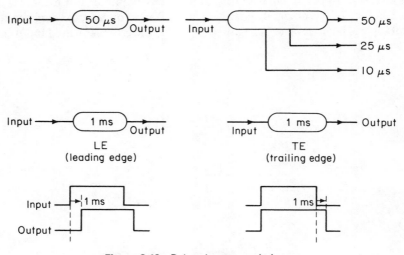

LE
(leading edge)

TE
(trailing edge)

Figure 2-18 Delay element symbols.

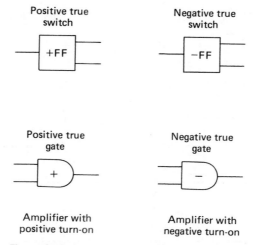

Figure 2-19   Truth polarity in digital logic symbols.

the polarity signs are a troubleshooting aid, indicating the polarity required to turn the amplifier on.

Inversion.   Generally, logic inversion is indicated by an inversion dot at inputs or outputs. (In some cases, inverted pulses are shown at the inputs and outputs). When the inversion dot appears on an input (generally only on gates and switching elements), the input is effective only when the input signal is of opposite polarity to that normally required. For example, if the FF in Fig. 2-20 is normally positive-true or is used on a digital diagram where all logic is positive-true, a negative-input (or the complemented input) at the

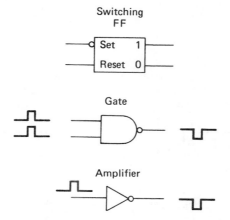

Figure 2-20   Methods of indicating inversion in digital logic symbols.

inversion dot will set the FF. (The terms "set" and "reset," as applied to FFs, are discussed in the following paragraphs.)

When the inversion dot appears at an output (generally only on gates and amplifiers), the output is of opposite polarity to that normally delivered. For example, if the gate of Fig. 2-20 is used in a positive-true logic circuit, the output will be negative. Similarly, the amplifier in Fig. 2-20 produces a positive output if the input is negative, and vice versa.

A-c coupling. Capacitor inputs to digital elements are often indicated by an arrow, as shown in Fig. 2-21. In the case of gates and switching elements, the element responds only to a change of the a-c coupled input in the true-going direction. An inversion dot used in conjunction with the coupling arrow indicates that the element responds to a change in the false-going direction. In the case of an amplifier, a pulse edge of the same polarity as given in the symbol turns the amplifier on briefly, then off as the capacitor charges. The output is then a pulse of the same width as the amplifier on-time. With an inversion dot at the amplifier output, the output pulse is inverted.

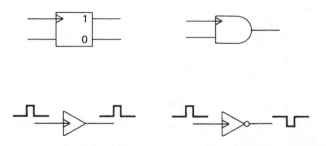

Figure 2-21 Methods of indicating a-c coupling in digital logic symbols.

Reference designations. Most digital diagrams show logic elements as a complete component rather than as the many components that make up the element. For example, an amplifier is shown as a triangle rather than as several dozen resistors, capacitors, transistors, and so on. Therefore, the amplifier has a reference designation of its own. On some digital diagrams, the logic-element symbols are mixed with symbols of individual resistors, capacitors, and so on. Either way, the logic-element symbol must be identified by a reference designation (to match descriptions in text or as a basis for parts listing).

The following logic-symbol reference designations are in general use: gates G, amplifiers and inverters A, flip-flops FF, one-shot OS (sometimes SS for single shot), multivibrator MV, Schmitt trigger ST, and delay D. Figure 2-22 shows some examples of how the reference designations are used. Note that the reference designations for switching elements are generally

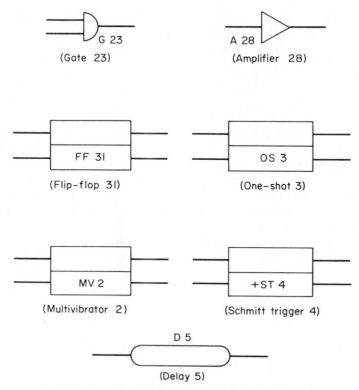

Figure 2-22   Reference designations for digital logic symbols.

placed within the symbol. All other designations are (usually) placed beside the symbol. In the case of switching elements, the true-state sign can be used as a prefix to the designation.

IC reference designations.   When digital logic elements appear in IC form (also called *microcircuit* form on some older digital diagrams), the system of reference designations is usually changed. Typical examples of IC logic reference designations are shown in Fig. 2-23.

When an IC is used to form one complete logic element, such as an amplifier or gate, the element can assume the IC reference designation of IC rather than G or A. When an IC is used to form a complete switching circuit, the element can assume the IC designator of IC, but it should also include the appropriate abbreviation (such as FF, OS, etc.) to identify its function. When a switching element is composed of portions of different IC packages, the designator IC of both packages should be included inside the symbol,

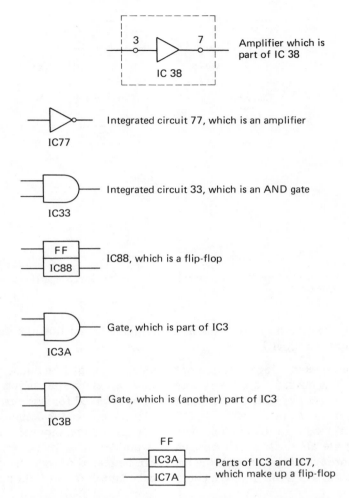

Figure 2-23 Reference designations for IC digital logic symbols.

and the appropriate identifying abbreviation should be located outside the symbol.

When more than one logic element is included in the IC package, which is the usual case, each logic element can be identified with a suffix: A, B, C, and so on. On some digital diagrams, the logic element is shown enclosed by dashed lines with the terminals identified. An example of this is shown in Fig. 2-23, where an amplifier symbol is enclosed by dashed lines and identified as IC38. This indicates that the amplifier is part of IC38 (integrated cir-

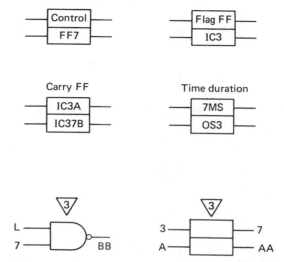

Figure 2-24   Reference names and location information on digital logic symbols.

cuit 38), that the input is available at terminal 3, and that the output is available at terminal 7.

Reference names.   Identification of digital elements by functional name in a diagram is normally done only with switching elements, as shown in Fig. 2-24. The upper portion of the symbol can be reserved for this purpose. If the type of switching element (FF, OS, etc.) is not part of the designation (in the lower portion of the symbol) then the reference name should include the appropriate abbreviation. If designations appear in both the upper and lower portions of the symbol, the name can be placed outside the symbol. For one-shots, it is often convenient to give the time duration as the reference name.

Location information.   Digital diagrams are intended to show the combination of logic elements that, taken together, form an instrument, part of an instrument, or a system of instruments. To aid in correlating the diagram with physical locations in instruments, additional information is given (by some manufacturers) with logic symbols as shown in Fig. 2-24. The number in the small triangle is sometimes used to indicate the circuit board on which the element appears. In the example of Fig. 2-24, the elements appear on circuit board 3. If all elements are on the same board, individual identification in this way is not necessary. Letters and numbers adjacent to inputs and outputs indicate pin numbers of the board (or the IC package), where inputs and/or outputs appear, as shown in Fig. 2-23.

2-8 IMPLEMENTING DIGITAL FUNCTIONS

The digital designer is often required to produce (or implement) a basic logic function using other logic functions. For example, a designer may be working with ICs that contain a number of general-purpose NAND gates. (Many manufacturers produce such ICs.) Although most of the design can be done using only NAND gates, assume that the design calls for the use of one inverter, or one NOR gate, and so on. It is possible to form all of the basic logic functions discussed thus far in this chapter by using only NAND gates.

Figure 2-25 shows the connections required to implement all of the basic logic functions, using other logic elements. For example, to form an inverter from a NOR gate, connect one input to ground or zero. To form the same inverter with a NAND gate, connect both inputs of the NAND gate together. With either connection, the output is inverted from the input.

2-8-2 Wired-AND and Wired-OR

When gate outputs are connected in parallel, and AND function or an OR function can result. These functions are often referred to as "wired-OR" and "wired-AND". If the true condition is represented by zero volts (or ground), then the parallel outputs or gates produce an OR function. If the true condition is represented by a voltage of any value or polarity, then the function is AND. The reason is that if a gate output is zero volts (ground), all the other outputs are, in effect, shorted to ground.

Any gates with active pull-up elements should not be connected for wired OR. When one gate has a 0 output and the other a 1 output, the resulting logic output is unpredictable.

2-9 DESIGNING DIGITAL CIRCUITS FROM SCRATCH

It is possible to buy such networks as decoders, encoders, flip-flops, counters, registers, arithmetic units, and memories in complete, functioning packages that require only a power source. A representative sampling of such packages is presented in Chapter 3. However, on the assumption that it may be necessary for you to design a basic circuit, the remaining sections of this chapter are devoted to logic circuit design "from scratch." A review of these sections will also help you understand the basics of logic design. We will start with some definitions.

Combinational digital circuits are those which have no feedback or memory (or do not depend upon feedback or memory for their operation). The combinational circuit produces an output (or outputs) in response to the presence of two or more variables (or inputs). The nature of the output

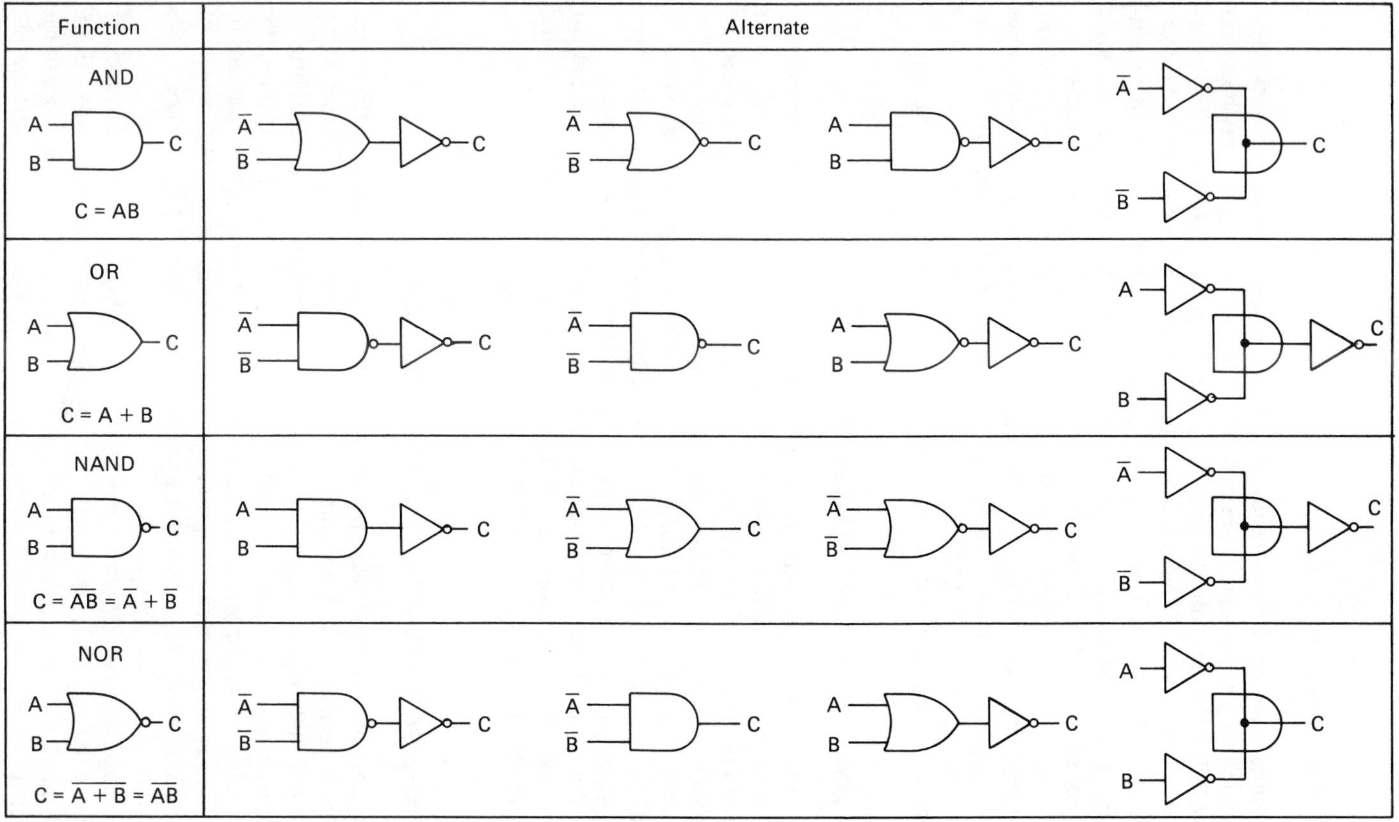

Figure 2-25 Implementing basic logic functions with other basic logic elements.

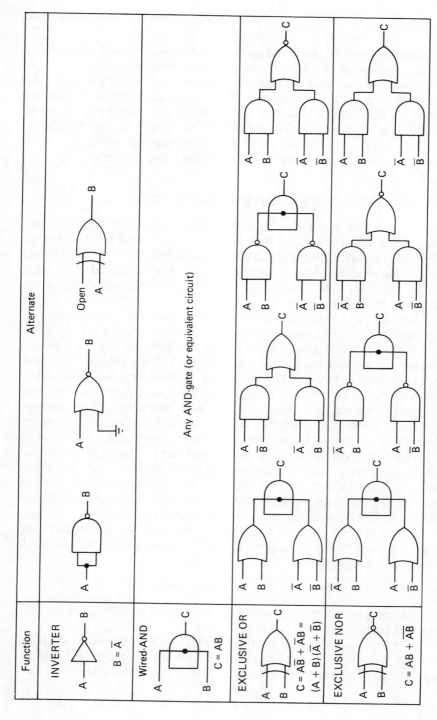

Figure 2-25 (*Continued*).

depends upon the combination of gates used. Adders, even and odd circuits, decoders, and encoders are examples of combinational circuits.

Sequential digital circuits have a memory, and possibly a feedback, and must operate in a given sequence. The outputs of sequential circuits depend on the time relation or timing of the inputs and the feedback, as well as the combination of gates or other elements. Sequential circuits usually involve flip-flops, and include such circuits as multivibrators, counters, registors, shift cells, arithmetic units, multipliers, and dividers.

2-10 DESIGNING BASIC COMBINATIONAL CIRCUITS

Problems in digital design may be stated in many ways. Similarly, you may receive the problem in many forms. For example, in some cases, you need only convert an existing equation (already simplified) into a working circuit. This process is often referred to as *implementing* or *mechanizing* an equation. In other cases, the equation must first be simplified, thus making the circuit equally simple.

Note that all equations should not be reduced to the simplest terms on an arbitrary basis. For example, an equation might contain three variables, one of which could be eliminated by proper manipulation of the equations. However, what if you must design a module that will solve the equation, but the module has three inputs? The third input cannot be ignored. In still other cases, you must write the equation from a truth table, or word statement, for the problem. No matter what the starting point, you should have a routine set of steps for the entire procedure. You should also use the same procedure repeatedly, until you no longer have to remember each step.

There are many procedures in common use by digital designers for writing, simplifying, and implementing equations. Some of these procedures involve the use of Veitch or Venn diagrams, or Karnaugh maps. The use of such diagrams and maps to simplify and implement digital circuits is discussed in Sec. 2-11. In this section, we use an even more basic procedure, based on simple truth tables and the rules of logical algebra. Once simplified to the desired terms, the equation can be implemented as a combinational network using basic logic gates.

2-10-1 Simplifying combinational equations

Commonly used Boolean equations are given in Fig. 2-3. The following rules and notes explain the equations.

The first (and probably most important) rule is: If two terms are identical except for one variable, and that variable is true in one term and inverted in the other term, the variable may be eliminated, and the two terms may be replaced by a common term. For example: $\overline{X}Y + XY$ can be replaced by Y, or $\overline{x}Y + xY = Y$.

For our purpose, a term can be considered as any group of variables, separated from other groups of variables or by the OR connective ($+$ sign). Thus, in the expression $AB\overline{C} + XY + ABC$, the $\overline{C}$ and C can be eliminated, and the two AB terms combined into one. The resulting expression is AB + XY.

A double negation converts an expression back to its original form. For example, $\overline{\overline{A}} = A$ and $XYZ = \overline{\overline{XYZ}}$.

The expression $\overline{A} + A$ is a true statement. This is because the expression indicates that $\overline{A}$ or A is to be considered.

The expression $\overline{A} \cdot A$ is false. That is, $\overline{A} \cdot A$ can never equal a true or 1. This is because the expression indicates the combination of a positive and a negative value, thus resulting in a cancellation.

The equation $A + (\overline{A}B) = A + B$, and can be so reduced in an equation.

The basic AND and OR functions are *communicative*. That is, the end results of the AND and OR connectives are not altered by the sequence that makes up the equation. For example, $A + B = B + A$.

Included in the communicative laws are De Morgan's laws: $\overline{AB} = \overline{A} + \overline{B}$ and $\overline{A} + \overline{B} = \overline{AB}$.

De Morgan's laws can be used to find the complement of any Boolean expression. For example, the negation of the AND ($\overline{AB}$) function (NOT-AND) is equal to the alternate denial ($\overline{A} + \overline{B}$), which expresses, in effect, that NOT-A or NOT-B is true.

In practical digital circuits, De Morgan's laws show the relationship of AND, OR, NAND, and NOR gates. If a NOT circuit follows an AND gate (the NAND function), the results are the same as inverting the A + B inputs to an OR gate. Similarly, if a NOT circuit follows an OR gate (the NOR function), it is comparable to the circuit obtained if the AB inputs to an AND gate are inverted.

The AND and OR functions are *associative*. The associative law states that the elements of a Boolean expression may be grouped as desired, as long as they are connected by the same sign. For example:

$$ABC = (AB)C = A(BC)$$
$$X + Y + Z = (X + Y) + Z = X + (Y + Z)$$

The AND and OR functions are also *distributive*. The distributive law has to do with the functional characteristics of logic connectives. If terms contain a common variable, even though the remainder of the terms are different, the common variables can be reduced to a single variable, and the terms can be recombined in simpler form, provided that the connective signs remain the same. For example:

$$(AB) + (AC) = A(B + C)$$
$$(A + B)(A + C) = A + (BC)$$

2-10-2 Product of sums versus sum of products

There are two classic solutions for the conversion of a truth table or word statement into a practical circuit: the *sum of products* (S-of-P) and the *product of sums* (P-of-S). Either solution can be applied to the same problem. This is shown in Fig. 2-26, where two circuits satisfy the conditions of the same truth table.

Although the circuits of Fig. 2-26 are workable, they have one obvious drawback. There is no amplification of the signals, since AND and OR gates are used. Then, since the digital pulses must pass through many gates or

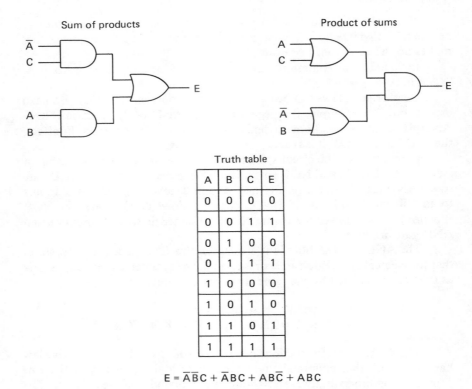

$$E = \overline{A}\,\overline{B}C + \overline{A}BC + AB\overline{C} + ABC$$

Figure 2-26 Comparison of sum-of-products and product-of-sums circuits for the same truth table and equation.

other circuits, it is common to provide some amplification, preferably at each gate, or at least for the group of gates that form a basic circuit.

This amplification is usually provided by the use of a NAND and NOR gate, or by AOI (AND-OR INVERT) networks. The S-of-P solution can best be implemented with NAND gates, whereas the P-of-S solution uses OR gates. The AOI solution requires AND gates, followed by a NOR gate.

2-10-3   A basic digital circuit using S-of-P

Figure 2-27 shows a truth table and a basic S-of-P circuit (using AND and OR gates). The following steps describe the procedure for implementing the circuit, starting with the word statement.

A typical word statement of the problem could be: "The circuit must accept three variables (or inputs); the circuit output must be false when variables ABC are false, and when AB are true and C is false, the circuit output must be true with all other combinations of variables."

To convert a word statement into a truth table, make sure that each variable has a separate column in the table. Since there are three variables in this problem, the truth table must have three columns. There are always eight rows in a three-variable truth table, so each variable can be in only one of two states. It is convenient to arrange the truth table the same way each time. Let the top row (or row number 1) have all zeros and the bottom row have all ones. Then let the second row from the top (row 2) shows a binary count of 1 (001), the third row (row 3) shows a binary count of 2 (010), and so on. The count of the next-to-bottom row (row 7 in this case) should be one less than the bottom row. In our case, the bottom row has a binary count of 7, whereas the next-to-last row has a count of 6, which is correct. In this way, the truth table will be consistent, and the number of rows will be correct (all possible combinations of variables will be present).

In an S-of-P solution, only those rows that produce a true (1) output need be considered. In this case, rows 1 and 7 can be omitted, since they produce a false output. An equation can be written using the remaining rows, as shown in Fig. 2-27. Then a basic digital circuit can be implemented from the equation. The circuit would require a separate AND gate for each of the six terms. The outputs of the AND gate would then be connected to a six-input OR gate.

However, it is possible to simplify the equation by using the steps shown in Fig. 2-27. These steps are based on the laws and rules of Sec. 2-10-1, and should be self-explanatory. As shown in Fig. 2-27, the final expression is reduced down to $\overline{A}B + A\overline{B} + C$. With the equation reduced to the shortest desired terms, the equation is implemented as follows:

If a term has only one variable, that variable (or input) is connected directly to the OR gate.

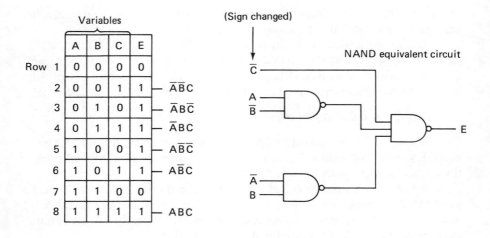

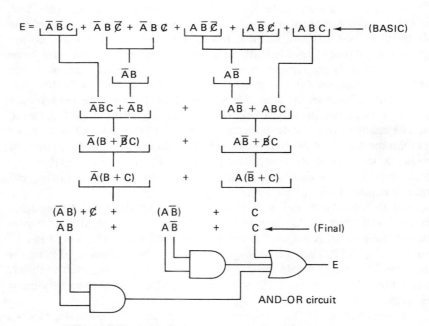

Figure 2-27 Basic S-of-P circuit (from truth table to circuit).

If a term has more than one variable, all variables of the term are connected to the input of an AND gate. In turn, the AND gate outputs are connected to the OR gate input.

Converting to NAND.   An S-of-P solution can be converted directly to a NAND network. That is, the OR gate and AND gates of an S-of-P network are replaced by NAND gates, as shown in Fig. 2-27. However, if the S-of-P circuit includes a single-variable term connected to the OR gate, the sign of that variable must be changed. This applies only to single-variable terms. For example, in Fig. 2-27, the single-variable C must be changed to $\bar{C}$.

2-10-4   A basic digital circuit using P-of-S

Figure 2-28 shows a truth table and a basic P-of-S circuit (using OR and AND gates). The word statement and truth table for this problem are identical to those of Fig. 2-27. However, in a P-of-S solution, only those rows of the truth that produce a false (0) output need be considered. In this case rows 1 and 7 only need be used. All other rows can be ignored.

The basic equation written from rows 1 and 7 is shown in Fig. 2-28. Note that the output of this equation is inverted (or false), since the false outputs are selected. To convert the output to a true, invert *both sides of the equation* (both sides of the equal sign). A double inversion of the output makes the output true. An inversion of the terms changes them from two AND terms, separated by an OR connective, into two OR terms, separated by an AND connective. With the equation reduced and changed, the equation is implemented as follows:

If a term has only one variable, that variable (or input) is connected directly to the AND gate input.

If a term has more than one variable, all variables of the term are connected to the input of an OR gate. In turn, the OR gate outputs are connected to the AND gate input.

Converting to NOR.   A P-of-S solution can be converted directly to a NOR network. That is, the AND gate and OR gates of the P-of-S network are replaced by NOR gates, as shown in Fig. 2-28. If the P-of-S circuit includes a single-variable term connected to the AND gate, the sign of that variable must be changed back to its original form. This applies only to single-variable terms (which do not occur in the circuit of Fig. 2-28).

2-10-5   A basic digital circuit using AOI

The AOI solution is a network consisting of AND gates followed by a NOR gate, and can be applied to the same digital circuits as those covered by S-of-P and P-of-S. As shown in Fig. 2-29, there are two methods of arriving

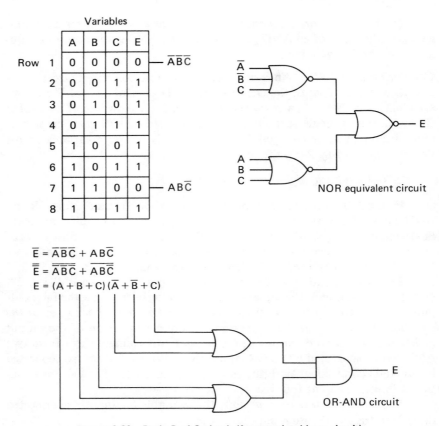

Figure 2-28   Basic P-of-S circuit (from truth table to circuit).

at an AOI network. Note that the truth table for each circuit of Fig. 2-29 is the same as the corresponding truth tables in Figs. 2-27 and 2-28.

When AOI is used to solve an S-of-P problem (using the variables that produce a true output), the equation is simplified and the circuit is implemented as follows:

If a term has more than one variable, all variables of the term are connected to the input of an AND gate. In turn, the AND gate outputs are connected to a NOR gate input.

If a term has only one variable, that variable is connected directly to the NOR gate input.

The output of the NOR gate represents an inversion of the final equation. This output must be inverted to satisfy the truth table. Any form of inverter can be used.

When AOI is used to solve a P-of-S problem (using the variables that

AOI solution (with and without inverted output)

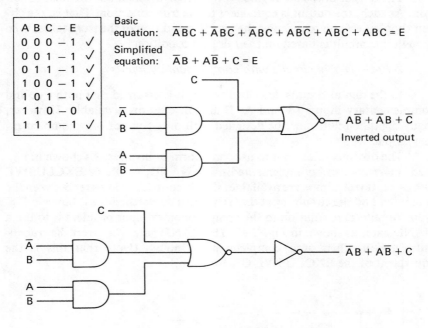

Basic
equation: $\overline{A}\,\overline{B}C + \overline{A}B\overline{C} + \overline{A}BC + A\overline{B}\,\overline{C} + A\overline{B}C + ABC = E$

Simplified
equation: $\overline{A}B + A\overline{B} + C = E$

A B C — E	
0 0 0 — 1	✓
0 0 1 — 1	✓
0 1 1 — 1	✓
1 0 0 — 1	✓
1 0 1 — 1	✓
1 1 0 — 0	
1 1 1 — 1	✓

$A\overline{B} + \overline{A}B + \overline{C}$

Inverted output

$\overline{A}B + A\overline{B} + C$

Alternate AOI solution (based on P-of-S)

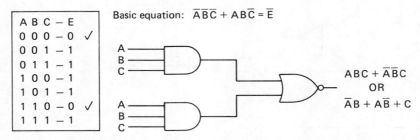

Basic equation: $\overline{A}\,\overline{B}\,\overline{C} + AB\overline{C} = \overline{E}$

A B C — E	
0 0 0 — 0	✓
0 0 1 — 1	
0 1 1 — 1	
1 0 0 — 1	
1 0 1 — 1	
1 1 0 — 0	✓
1 1 1 — 1	

$ABC + \overline{A}\,\overline{B}C$
OR
$\overline{A}B + A\overline{B} + C$

Figure 2-29 Basic AOI circuit (from truth table to circuit).

produce a false output), the equation is simplified and the circuit is implemented as follows:

If a term has more than one variable, all variables of the term are connected to the input of an AND gate. In turn, the AND gate outputs are connected to a NOR gate input.

If the term has only one variable, that variable is connected directly to the NOR gate input.

The output of the NOR gate represents an inversion of the false condition. As such, the output is equivalent to the true condition. That is, the circuit should perform to produce the true (1) condition of the truth table, even though the circuit is based on the false (0) condition.

2-10-6 Digital circuits with complementary inputs

In the digital circuits described so far, it is assumed that both true and complementary inputs (A and A, B and B, etc.) are available. However, many digital circuits must be designed with only true (or only complementary) inputs.

The obvious solution is to use some form of inverter. As shown in Fig. 2-25, inverters can be implemented using NAND, NOR, or EXCLUSIVE OR gates. If true inputs are available, simply connect an inverter between the true input and the circuit input that requires a complement, as shown in Fig. 2-30. An alternate solution to the complementary input problem is to use a NAND gate, as shown in Fig. 2-30. The NAND gate will invert the true inputs, making them into complementary inputs. (Note that this is the equivalent of the EXCLUSIVE OR function.)

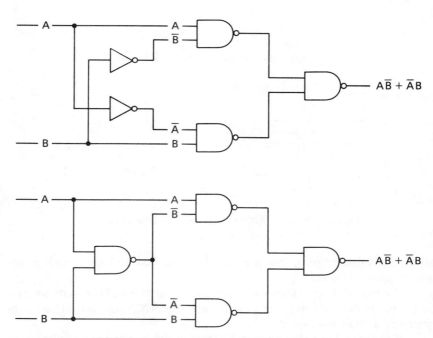

Figure 2-30 Providing complementary inputs by means of inverters and NAND gates.

2-10-7 Some basic digital terms

The terms *fan-in, fan-out, load,* and *drive* are used interchangeably in digital literature. To avoid confusion, the following definitions apply to digital circuits described in this book.

The term *fan-in* is generally applied to digital circuits in IC form. Fan-in is defined here as the number of independent inputs on a digital element. For example, a three-input NAND gate has a fan-in of 3.

Each input to a digital element represents a *load* to the circuit trying to drive the element. If there are several digital elements (say several gates) in a given circuit, the load is increased by one unit for each element. Load, therefore, means the number of elements being driven from a given input.

Fan-out (primarily an IC term) is defined as the number of digital elements that can be driven directly from an output, without any further amplification or circuit modification. For example, the output of a NAND gate with a fan-out of three can be connected directly to three elements. The term *drive* can be used in place of fan-out.

2-10-8 Digital delay

Any solid-state element (diode, transistor, and so on) will offer some delay to signals or pulses. That is, the output pulse will occur some time after the input pulse. Thus, every digital element has some delay. This delay is known as *propagation delay time, delay time,* or simply *delay.* Some digital literature spells out a specific time (usually in nanoseconds). Other digital literature specifies delay for a given circuit as the number of elements between a given input and given output. For example, a delay of three units is presented by the circuits of Fig. 2-30.

The problems presented by delay become obvious when it is realized that most digital circuits operate on the basis of coincidence. For example, an AND gate produces a true output only when both inputs are true, simultaneously. If input A is true, but switches back to false before input B arrives (due to some delay of the B input), the inputs do not occur simultaneously and the AND gate output is false. The problem is compounded when delay occurs in sequential networks, which depends upon timing as well as coincidence. Many of the failures that occur in digital circuits are the result of undesired delay.

2-11 DIGITAL MAPPING

Digital literature often uses mapping to show operation of circuits. The Venn and Veitch diagrams, as well as Karnaugh maps, are the most used digital mapping tools. This section summarizes these tools. A comprehensive discussion of digital mapping is contained in the author's best selling *Logic Designer's Manual* (Reston, Va.: Reston Publishing Company, 1977).

2-11-1   Venn diagrams

The Venn diagram is a pictorial representation of logical expressions. Figure 2-31 shows some typical examples of Venn diagrams for basic equations and logic functions.

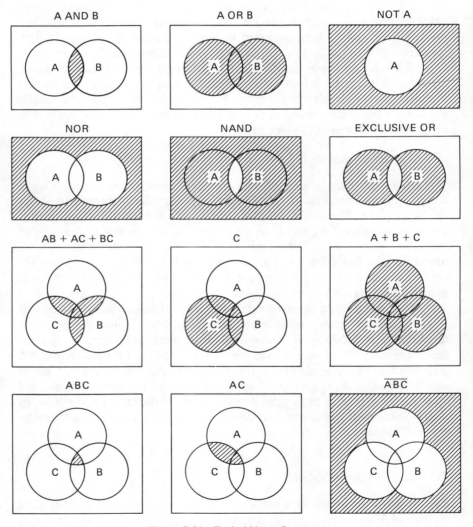

Figure 2-31   Typical Venn diagrams.

2-11-2 Veitch diagrams

Veitch diagrams provide a graphic means of representing logic equations. In effect, a Veitch diagram is a from of truth table that can be used to represent the terms of the equation, and to simplify the equation. We will not go into the simplification procedure here, since Veitch diagrams have generally been replaced by Karnaugh maps. However, you should be able to convert an equation into a Veitch diagram, and vice versa.

Figure 2-32 shows a four-variable equation and the corresponding Veitch diagram. (A Veitch diagram can have any number of variables, but six variables is usually the practical limit.) Since each variable has two possible states (true or false), the number of squares required in a Veitch diagram is 2^n, where n is the number of variables. Thus, for the four-variable equation of Fig. 2-32 (ABCD), there are 16 squares in the Veitch diagram.

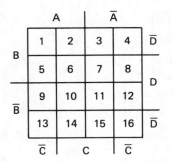 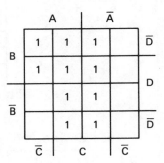

$$F = AB + \overline{A}BCD + A\overline{B}C + \overline{A}C + ACD$$

Figure 2-32 Example of Ve. ch diagram for four variables.

The equation is plotted on the Veitch diagram by placing a 1 in each square that is represented by the terms of the equation. For example, in the equation of Fig. 2-32, a 1 is placed in each square as follows:

$$
\begin{aligned}
F = &\ AB && \text{squares 1, 2, 5, and 6}\\
+ &\ \overline{A}BCD && \text{square 7}\\
+ &\ A\overline{B}C && \text{squares 10 and 14}\\
+ &\ \overline{A}C && \text{squares 3, 7, 11, and 15}\\
+ &\ ACD && \text{squares 6 and 10}
\end{aligned}
$$

2-11-3 Karnaugh maps

The Karnaugh map is a digital logic tool similar to the Veitch diagram in that Karnaugh maps can substitute for a truth table or be used to simplify an equation. The Karnaugh map can also be used to convert logic equations

Truth table

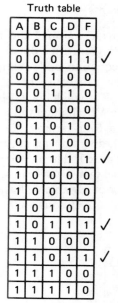

A	B	C	D	F	
0	0	0	0	0	
0	0	0	1	1	✓
0	0	1	0	0	
0	0	1	1	0	
0	1	0	0	0	
0	1	0	1	0	
0	1	1	0	0	
0	1	1	1	1	✓
1	0	0	0	0	
1	0	0	1	0	
1	0	1	0	0	
1	0	1	1	1	✓
1	1	0	0	0	
1	1	0	1	1	✓
1	1	1	0	0	
1	1	1	1	0	

$$F = \overline{A}\,\overline{B}\,\overline{C}D + \overline{A}BCD + A\overline{B}CD + AB\overline{C}D$$

	$\overline{C}\overline{D}$	$\overline{C}D$	CD	$C\overline{D}$
$\overline{A}\overline{B}$		1		
$\overline{A}B$			1	
AB		1		
$A\overline{B}$			1	

C\D A B	00	01	11	10
00		1		
01			1	
11		1		
10			1	

Figure 2-33 Four-variable Karnaugh maps for truth table and equation.

or truth tables into corresponding digital circuit diagrams in a simple and orderly manner.

Figure 2-33 shows a four-variable Karnaugh map for a corresponding truth table and equation. Note that there are two alternate formats for the map. The author recognizes that many other formats exist for Karnaugh maps but has found the illustrated forms to be in most common use.

Note that each square of the Karnaugh map corresponds to a row of the truth table and to one term of the equation. For example, row 1 of the truth table (the term $\overline{A}\,\overline{B}\,\overline{C}D$) corresponds to the upper left-hand square of the Karnaugh map. Since the output of this row is 0 (false), the correspond-ing map square can contain a 0. However, when you are only interested in outputs, the zeros can be omitted from the map. Also, the squares can be marked with letters T (for true) or F (for false) if desired. No matter what scheme is used, the rows and columns must be labeled so that only one variable changes at a time as you move from square to square.

For example, in the lower left-hand square of the map $(\overline{A}\overline{B}\overline{C}\overline{D})$, if you move one square up, the A and $\overline{C}\overline{D}$ remain the same but the $\overline{B}$ changes to a B. Also, starting from the same lower left-hand square, if you move one square to the right, the $\overline{A}\overline{B}\overline{C}$ remain the same but the $\overline{D}$ changes to a D.

Simplifying the equation with a Karnaugh map.

The arrangement of adja-cent squares permits the Karnaugh map to be used to simplify equations and to convert equations or truth tables directly to circuit diagrams. The first step in the simplifying procedures is to draw loops around the adjacent 1s on the map. The following rules are illustrated in Fig. 2-34.

On a two-variable map, any two adjacent squares can be looped. On a three-variable map, any two or four adjacent squares can be looped. On a four-variable map, any two, four, or eight squares can be looped. The top row is considered to be adjacent to the bottom row, and the extreme right-hand row is adjacent to the extreme left-hand row. When four 1s are looped, they can be in one row, one column, or in a 2-by-2 square. When eight 1s are looped, they must be in a 2-by-4 square. Loops may overlap other loops, but each loop must be treated as a separate term of the equation. Any 1s not covered by adjacent loops must be looped separately.

There are many alternate methods for looping and there are no fixed rules as to the best method. However, as a general rule, always make the loops as large as possible by first drawing loops of eight, then four, then two. Small loops do not provide the simplest equations, nor do they produce the minimum solutions to circuit implementations.

Once the loops have been drawn on the map, each loop is converted to a term of the simplified equation, using the following general rule. When a variable appears in both true and complemented form within a loop, the variable can be eliminated. When the variable remains the same for all squares in the loop, that variable must be included in the simplified term of the equation. This is shown in Fig. 2-35 where there are two loops in a four-variable map.

In the four-in-a-row loop, the W and X variables remain constant (true only), while the YZ variables appear in both true and complemented forms. This results in a final term of WX. In the two-by-two loop, the X and Z variables appear in true-only form, while the WY variables appear in both true and complemented forms. This results in a final term of XZ. The com-bined terms result in a final simplified equation WX + XZ.

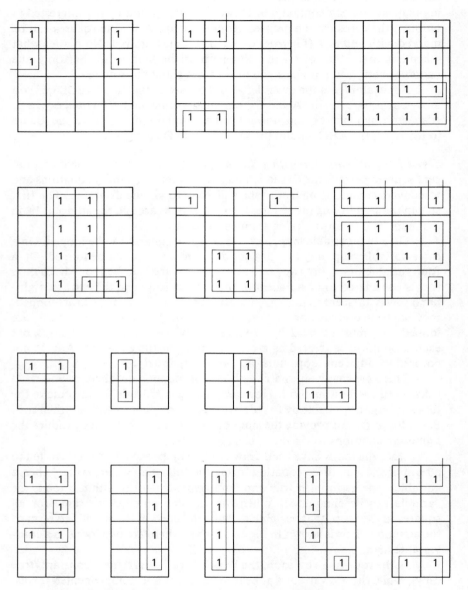

Figure 2-34 Looping of Karnaugh maps for simplification of equations.

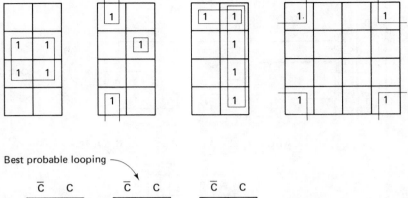

Best probable looping

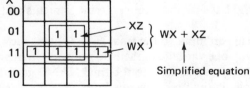

Figure 2-34 (*continued*).

Original equation: $\overline{W}X\overline{Y}Z + \overline{W}XYZ + WX\overline{Y}\overline{Z} + WX\overline{Y}Z + WXYZ + WXY\overline{Z}$

$$
\begin{array}{c|c|c|c|c}
 & 00 & 01 & 11 & 10 \\
\hline
00 & & & & \\
\hline
01 & & 1 & 1 & \\
\hline
11 & 1 & 1 & 1 & 1 \\
\hline
10 & & & & \\
\end{array}
$$

XZ $\Big\}$ WX + XZ
WX $\Big\}$

↑
Simplified equation

Figure 2-35 Using a Karnaugh map to simplify an equation.

Going from Karnaugh directly to a digital circuit.   It is not necessary to simplify equations and then convert the equations to a circuit diagram when working with Karnaugh maps. It is possible to go directly from the mapping to a circuit diagram. The following is an example of converting a Karnaugh map directly to a NAND circuit.

The first step is to loop all 1s (using the rules previously discussed), as shown in Fig. 2-36. Note that the truth table is the same as that used in Fig. 2-27. Three loops are drawn on the map of Fig. 2-36. More loops could have

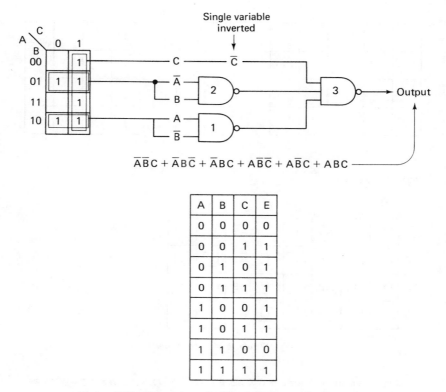

$$\overline{A}\overline{B}C + \overline{A}B\overline{C} + \overline{A}BC + A\overline{B}\overline{C} + A\overline{B}C + ABC$$

A	B	C	E
0	0	0	0
0	0	1	1
0	1	0	1
0	1	1	1
1	0	0	1
1	0	1	1
1	1	0	0
1	1	1	1

Figure 2-36   Converting a Karnaugh map to a NAND circuit.

been used, but this would have required more circuit elements and more inputs.

Each loop represents an input to the final NAND gate (shown as gate 3). If a loop contains only one variable in the true-only or complement-only form, that variable is connected directly to the final NAND gate, but in inverted form. For example, in the four-in-a-row loop, only the C variable remains constant (true-only), while AB appear in both true and complemented forms. This C is inverted to $\overline{C}$ and connected directly to the final NAND gate.

If a loop contains two or more variables that remain in the true only or complemented form, those variables are connected to a separate NAND gate. In turn, the output of the separate gate (or gates) is connected to the final NAND gate. For example, in the bottom loop, only A and B variables remain constant. Thus, A and $\overline{B}$ are connected to a separate NAND gate (gate 1), whereas the output of this gate is connected to the final NAND gate (gate 3). In the top loop, only the $\overline{A}$ and B variables remain constant. Thus,

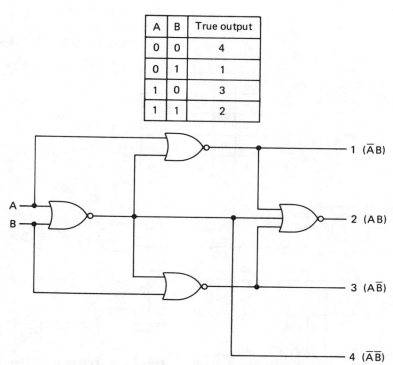

A	B	True output
0	0	4
0	1	1
1	0	3
1	1	2

1 ($\overline{A}B$)

2 (AB)

3 (A$\overline{B}$)

4 ($\overline{A}\overline{B}$)

Figure 2-37 Two-variable decoder using NOR gates.

$\overline{A}$ and B are connected to a separate NAND gate (gate 2), while this gate is connected to the final NAND gate (gate 3).

2-12 SOME TYPICAL COMBINATIONAL CIRCUITS

There are many types of combinational circuits used in digital electronics. Similarly, there are an infinite variety of combinational circuits available in IC form. The following paragraphs describe some typical examples.

2-12-1 Decoder circuits

The basic function of a combinational circuit is to produce an output (or outputs) only when certain inputs are present. Thus, combinational circuits indicate the presence of a given set of inputs by producing the corresponding output. Decoders are classic examples of combinational circuits. In effect, all combinational circuits are decoders of a sort.

Variable input decoders produce an output that indicates the state of input variables. For example, if it is assumed that each variable can have only two states (0 or 1), there are four possible combinations for two variables (00, 01, 10, and 11). Thus, a two-variable decoder has two inputs (one for each variable) and four outputs (one for each possible combination). Figure 2-37 shows the circuit of a two-variable decoder using NOR gates. As the

69

A	B	True output
0	0	1
0	1	3
1	0	2
1	1	4

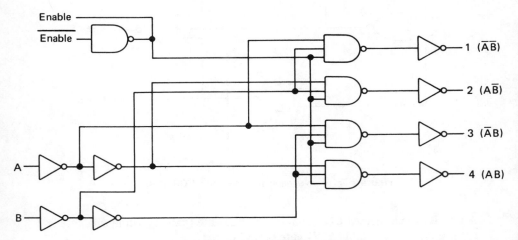

Figure 2-38 Gated two-variable decoder.

truth table shows, one and only one output is true for each of the four possible input states.

The circuit of Fig. 2-38 is a *gated* two-variable decoder. Such circuits are often found in IC form. The outputs are available only when an "enable," "strobe," "gate," or "clock" signal is present. Such "turn-on" signals can be in pulse form or can be a fixed d-c voltage controlled by a switch. Note that the enable signal can be either true or complemented.

Even and odd decoders indicate the number of variables that are true (or false) on an "even or odd" basis. For example, an "odd" decoder with three variables produces a true output only when an odd number (1 or 3) of the input variables is true. Such a decoder circuit is shown in Fig. 2-39.

Majority and minority decoders are used to indicate the majority (or minority) state of the input variables. For example, if a majority decoder such as shown in Fig. 2-40a is used with three variables, the single output is true only when two or three inputs are true. With the minority decoder of Fig. 2-40b, the output is true if any one of the inputs does not agree with the other two inputs. If all three inputs agree, either true or false, the output is false.

A	B	C	Output
0	0	0	0
0	0	1	1
0	1	0	1
0	1	1	0
1	0	0	1
1	0	1	0
1	1	0	0
1	1	1	1

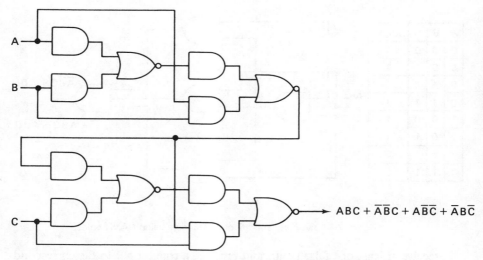

$$ABC + \overline{A}\,\overline{B}C + A\overline{B}\,\overline{C} + \overline{A}B\overline{C}$$

Figure 2-39 Three-variable odd decoder using AOI gates.

Code converters are a very common type of decoder found in IC form. Such decoders convert one type of digital code to another. For example, a binary-to-decimal decoder will convert a 4-bit binary number into a decimal equivalent. The circuit of Fig. 2-41 is a BCD-to-decimal decoder. With this decoder one, and only one, output is true when the corresponding inputs are true. For example, the BCD equivalent of decimal 7 is 0111. Thus, the MSB of the binary input is false, whereas the complement of the MSB input is true. The remaining 3 bits of the binary inputs are all true (with their complements false). Gate 7A receives three inputs (lines A, B, and C) and produces a corresponding false output to gate 7B. The output of gate 7B is inverted to true (7B is a NAND gate used as an inverter). All other A gates

A	B	C	Output
0	0	0	0
0	0	1	0
0	1	0	0
0	1	1	1
1	0	0	0
1	0	1	1
1	1	0	1
1	1	1	1

A	B	C	Output
0	0	0	0
0	0	1	1
0	1	0	1
0	1	1	1
1	0	0	1
1	0	1	1
1	1	0	1
1	1	1	0

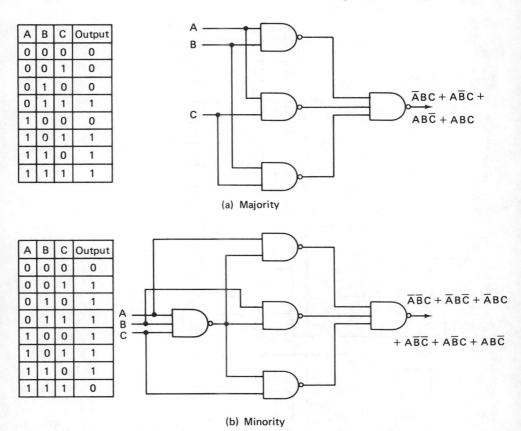

(a) Majority

(b) Minority

Figure 2-40   Majority and minority decoders using NAND gates.

receive at least one false input, and produce a true output to the corresponding B (or inverter) gate. For example, gate 3A receives two true inputs (lines
A and B) and one false input (line $\overline{C}$). Thus, gate 3A produces a true output,
which is inverted to a false by gate 3B.

A typical line of digital IC code converters could include binary or
BDC to octal, BCD to hex, binary to ASCII, and so on. Figure 2-42 shows a
Texas Instruments SN54/741ɔ4 IC decoder. This is a classic device of the TI
54/74 series. (The designation 54 applies to military, whereas 74 applies to
industrial IC digital devices manufactured by TI.) Note that Fig. 2-42 shows
the truth table and block diagram of the IC rather than the complete circuit.
This block representation is typical of the diagrams found in digital electronic literature.

As shown in Fig. 2-42, the device is housed in a 24-pin IC. Power is applied to pins 12 and 24. The remaining pins are for active signals. (Digital ICs

Decimal	BCD			
	D	C	B	A
0	0	0	0	0
1	0	0	0	1
2	0	0	1	0
3	0	0	1	1
4	0	1	0	0
5	0	1	0	1
6	0	1	1	0
7	0	1	1	1
8	1	0	0	0
9	1	0	0	1

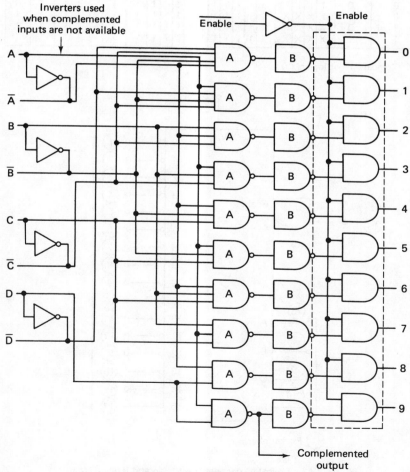

Figure 2-41 BCD-to-decimal decoder.

Inputs						Outputs															
Enable	Data	D	C	B	A	0	1	2	3	4	5	6	7	8	9	10	11	12	13	14	15
0	0	0	0	0	0	0	1	1	1	1	1	1	1	1	1	1	1	1	1	1	1
0	0	0	0	0	1	1	0	1	1	1	1	1	1	1	1	1	1	1	1	1	1
0	0	0	0	1	0	1	1	0	1	1	1	1	1	1	1	1	1	1	1	1	1
0	0	0	0	1	1	1	1	1	0	1	1	1	1	1	1	1	1	1	1	1	1
0	0	0	1	0	0	1	1	1	1	0	1	1	1	1	1	1	1	1	1	1	1
0	0	0	1	0	1	1	1	1	1	1	0	1	1	1	1	1	1	1	1	1	1
0	0	0	1	1	0	1	1	1	1	1	1	0	1	1	1	1	1	1	1	1	1
0	0	0	1	1	1	1	1	1	1	1	1	1	0	1	1	1	1	1	1	1	1
0	0	1	0	0	0	1	1	1	1	1	1	1	1	0	1	1	1	1	1	1	1
0	0	1	0	0	1	1	1	1	1	1	1	1	1	1	0	1	1	1	1	1	1
0	0	1	0	1	0	1	1	1	1	1	1	1	1	1	1	0	1	1	1	1	1
0	0	1	0	1	1	1	1	1	1	1	1	1	1	1	1	1	0	1	1	1	1
0	0	1	1	0	0	1	1	1	1	1	1	1	1	1	1	1	1	0	1	1	1
0	0	1	1	0	1	1	1	1	1	1	1	1	1	1	1	1	1	1	0	1	1
0	0	1	1	1	0	1	1	1	1	1	1	1	1	1	1	1	1	1	1	0	1
0	0	1	1	1	1	1	1	1	1	1	1	1	1	1	1	1	1	1	1	1	0
0	1	X	X	X	X	1	1	1	1	1	1	1	1	1	1	1	1	1	1	1	1
1	0	X	X	X	X	1	1	1	1	1	1	1	1	1	1	1	1	1	1	1	1
1	1	X	X	X	X	1	1	1	1	1	1	1	1	1	1	1	1	1	1	1	1

X = logical "1" or logical "0"

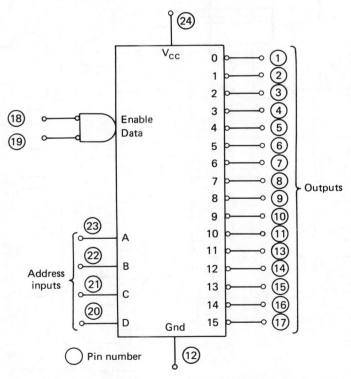

Figure 2-42 Typical IC decoder truth table and block diagram.

are discussed further in Chapter 3.) The four binary-coded address inputs A, B, C, and D are decoded to select (or *address*) only one of 16 mutually exclusive outputs. All 16 outputs are at 1 unless both enable and data inputs (pins 18 and 19) are at 0. When enable and data inputs are 0, the addressed output is at 0, with all other outputs at 1. For example, if A, B, and C are at 0 and D is 1 (binary 1000 or decimal 8), then the address output is 8 (8 is at 0, all other outputs are at 1).

The device can also be used as a 1-line to 16-line data distributor by setting enable at 0 and connecting binary information to the data input. (Data distributors are discussed further in Sec. 3-12-4.) This information is routed, unchanged, to the output selected by the address inputs. For example, if A and B are at 1, C and D are at 0, and enable is at 0, whatever data pulses appear on the data input will be distributed to output 3. All other outputs remain at 1.

2-12-2 Encoder circuits

The term "encoder" is used here to indicate any combinational circuit that provides the opposite function of a decoder. For example, the decoder of Fig. 2-41 converts a 4-bit BCD number to a 10-bit decimal number. An encoder, in this context, converts a 10-bit decimal number to its BCD equivalent. However, both encoders and decoders are forms of combinational circuits in that they both produce an output (or output) only when certain inputs are present.

The circuit of Fig. 2-43 is a decimal-to-BCD encoder. Operation of the circuit is as follows. Assume that the three of the decimal inputs are true and that all other inputs are false. (Only one of the inputs can be true at the same time.) Under these conditions, the output of inverter gate 3 is false, while the output of all other inverter gates is true. Inverter gate 3 is connected to gates 1 and 2. Any false input to a NAND gate produces a true output. Thus, output gates 1 and 2 are true. The inputs to output gates 4 and 8 are all true. All true inputs to a NAND gate produce a false. Thus output gates 4 and 8 are false. The binary output is then 0011, or decimal 3.

Note that the 0 decimal input line is not connected. When the decimal input is 0, lines 1 through 9 are all false, the outputs of the inverter gates are all true, and the output gates are all false, producing a binary 0000, or decimal 0.

2-12-3 Parity and comparator circuits

In digital systems special codes have been specifically designed for detecting errors that might occur in digital counting. Any complex digital network that operates on a binary counting system is subject to counting er-

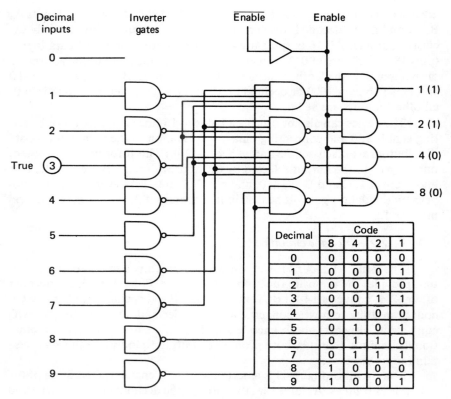

Figure 2-43   Decimal-to-BCD encoder.

rors due to circuit failure, noise on the transmission line (say, over the telephone lines connected between computer terminals), or similar occurrence. For example, a defective amplifier can reduce the amplitude of a pulse (representing a binary 1) so that it appears as a binary 0 at the following circuit. Similarly, noise in a circuit can be of equal amplitude to a normal binary 1 pulse. If the noise occurs at a time when a particular circuit is supposed to receive a binary 0, the circuit can react as if a binary 1 is present.

Parity check.   A parity check is one method for detecting errors in digital circuits. Other methods are discussed in Chapters 3 and 4. Parity refers to the quality of being equal, and a parity check is a equality-checking code. The coding consists of introducing additional bits or pulses into the binary number. The additional bit is known as a *parity bit,* and may be either a 0 or a 1. The parity bit is chosen to make the number of all bits in the binary group even or odd. If a system is chosen in which the bits in the binary number, *plus the parity bit,* are even, the system is known as even parity. Even parity is in general use. However, odd parity can be used.

As an example of even parity, 7F in hex or 0111-1111 in binary requires a 1 as the parity bit. Thus, in the total of 8 bits (7 binary bits, plus the one parity bit, or 1111-1111), there is an even number of 1s.

Error-detection circuits based on the parity system use both *parity generators* and *parity detectors* or *checkers*. The function of a parity generator is to examine the word (or group of binary bits) and calculate the information required to for the added parity bit. For example, if there are three 1s in the binary group and even parity is used, a 1 must be used for the parity bit to make an even number in the "parity word."

Once the parity bit has been included (a pulse has been added on the appropriate line of the data bus), the parity word (binary bits, plus parity bit) can be examined after any transmission, or at any point in the system, to determine if a failure or error has occurred. A parity detection circuit or parity checker examines the parity word to see if the desired odd or even parity still exists (say after passing through several ICs in a digital system). If an error has occurred, the system control can be informed that the system is not functioning properly by means of a signal. Parity circuits are often used with peripheral equipment such as tape recorders and magnetic disks. For that reason, parity circuits are discussed further in Chapter 4.

Simple parity checking system. The fundamental operation required in parity generation and detection circuits, that of comparing inputs to determine the presence of an odd or even number of 1s, can be effectively performed by EXCLUSIVE OR gates. The basic EXCLUSIVE OR gate performing the function $\overline{A}B + A\overline{B}$ serves to calculate parity over inputs A and B. EXCLUSIVE OR (and EXCLUSIVE NOR) gates can be interconnected to form *parity trees,* and thus perform parity checks on longer words. Figure 2-44 shows the circuits of a 4-bit and an 8-bit parity tree. The 8-bit parity tree consists of seven EXCLUSIVE NOR gates. Each gate has a 0 output if, and only if, one of its two inputs is at a 1 level. Thus, the output of the three-stage parity tree is in the 0 state if there is an odd number of ones over the eight inputs. The 4-bit parity tree produces a 0 output if an odd number of 1s exists at the four inputs.

The circuits of Fig. 2-44 form the basic building blocks for a 20-bit word simple even-parity generator and detector shown in Fig. 2-45. Such a system detects the presence of a single error in a word. If two errors occur, the output does not indicate that an error has occurred. Thus, simple parity will detect an odd number of errors but will fail if an even number of errors occurs.

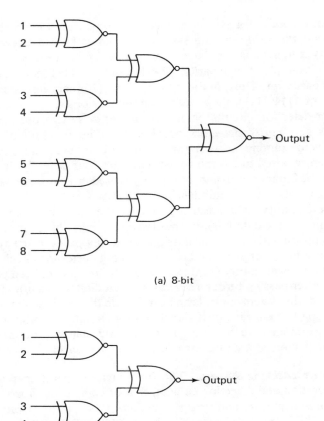

(a) 8-bit

(b) 4-bit

Figure 2-44 Eight-bit and 4-bit parity trees using EXCLUSIVE NOR gates.

Advanced parity systems.   Parity-checking systems have been devised to recognize that an error has occurred, and to detect *which bit is in error.* Several extra bits must be added to the system word to make these parity checks. One such system is referred to as Hamming parity single-error detection and correction (after R. W. Hamming). We will not go into the full details of this or any other parity-checking system here. However, the author's *Logic Designer's Manual* referenced in Sec. 2-11 does cover full circuit details of the Hamming system for a 16-bit message word.

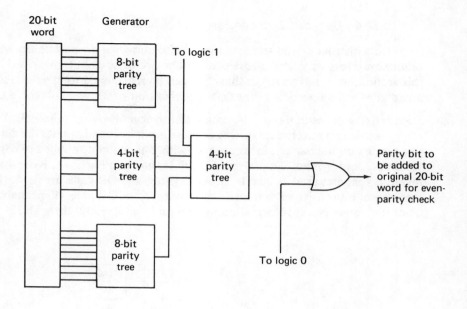

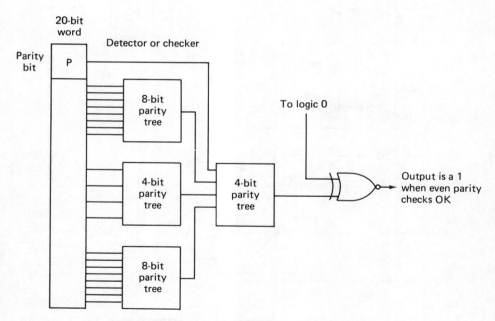

Figure 2-45 Basic even-parity generator and detector circuits.

2-12-4 Data distributor and selector circuits

Data distributor and selector circuits (also known as *demultiplexers* and *multiplexers*) are very similar to decoders. The two basic circuits described in this section are "universal" in that they can be rearranged to meet almost any digital system needed where data distribution or selection is involved.

Data selector or multiplexer. A data selector or multiplexer (also called a *multiplex switch*) selects data on one or more input lines and applies the data to a single output channel, in accordance with a binary code applied to control lines. A four-input data selector circuit is shown in Fig. 2-46. Note that inverters (or gates used as inverters), AND gates, and OR gates are used. (If the selector must drive several loads at the output, the OR gate will probably be of the power type.) Information present on input lines X0 through X3 is

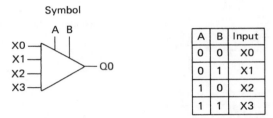

A	B	Input
0	0	X0
0	1	X1
1	0	X2
1	1	X3

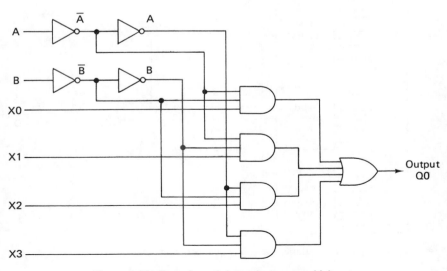

Figure 2-46 Four-channel data selector or multiplexer.

C	D	A	B	Q0
0	0	0	0	X0
0	0	0	1	X1
0	0	1	0	X2
0	0	1	1	X3
0	1	0	0	X4
0	1	0	1	X5
0	1	1	0	X6
0	1	1	1	X7
1	0	0	0	X8
1	0	0	1	X9
1	0	1	0	X10
1	0	1	1	X11
1	1	0	0	X12
1	1	0	1	X13
1	1	1	0	X14
1	1	1	1	X15

One bit from one of 16 locations

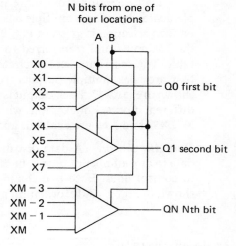

N bits from one of four locations

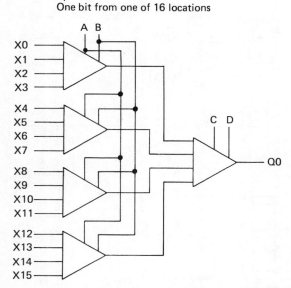

Figure 2-47 Data selectors with expanded selection capability.

transferred to output Q0, in accordance with the state of control inputs A and B, as shown by the truth table.

Data selection from more than four locations (or inputs) can be implemented by using a multiple basic data selector circuits. An example of expanded selection is given in Fig. 2-47. Here, one bit is selected from one of the 16 inputs and transferred to the output in accordance with the truth table. The basic data selector circuit can also be used to implement a network that provides multiple outputs from multiple inputs. Such a circuit is also shown in Fig. 2-47. Each output is selected from one of its own group of four different inputs. For example, with control lines A and B both at 1, data lines X3, X7, and XM are enabled, and all other data lines are turned off.

Data distributor. A data distributor distributes a single channel of input data to any number of output lines, in accordance with a binary code applied to control lines. Both two-channel and four-channel data distributors are shown in Fig. 2-48. Information applied to input X is distributed to outputs

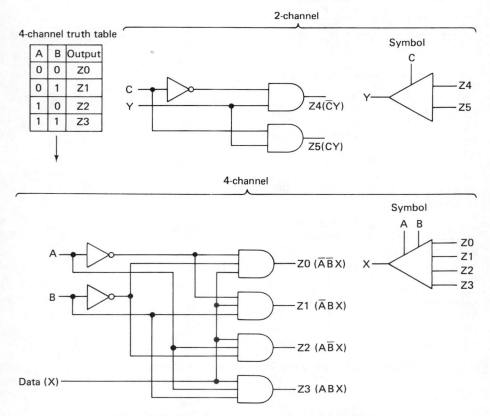

Figure 2-48 Two-channel and four-channel data distributors.

B	A	C	D	Output
0	0	0	0	1
1	0	0	0	2
0	1	0	0	3
1	1	0	0	4
0	0	1	0	5
1	0	1	0	6
0	1	1	0	7
1	1	1	0	8
0	0	0	1	9
1	0	0	1	10
0	1	0	1	11
1	1	0	1	12
0	0	1	1	13
1	0	1	1	14
0	1	1	1	15
1	1	1	1	16

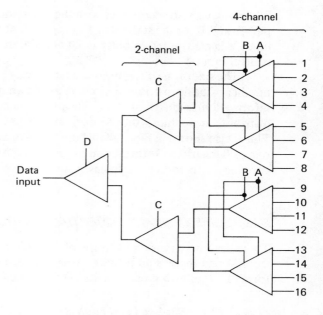

N bits, each bit to one of 8 locations

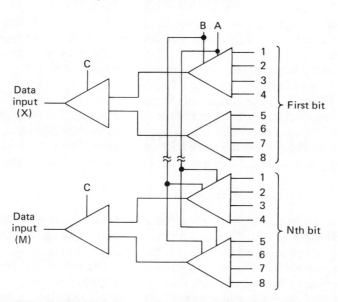

Figure 2-49   Data distributors with expanded distribution capability.

83

Z0 through Z3, in accordance with the binary numbers applied to control inputs A and B (or the states 0 or 1 of inputs A and B). Information applied to input Y is distributed to outputs Z4 and Z5, in accordance with the state of control input Y.

The size of the distribution function may be increased by increasing the number of basic distributor circuits, as shown in Fig. 2-49, where one bit of information is distributed to one of 16 locations. It is also possible to distribute more than 1 bit of information using the basic distributor circuit. This is also shown in Fig. 2-49, where N bits of information are distributed to one of 8 locations. Information on inputs X through M is distributed in accordance with control variables A and B.

2-13   SOME TYPICAL SEQUENTIAL CIRCUITS

There are many types and forms of sequential circuits used in digital electronics, and there is an infinite variety of sequential circuits available in IC form. The following paragraphs describe some typical examples.

2-13-1   Flip-flops and latches

Sequential circuits are based on the use of flip-flops (FFs) and latches. (A latch is essentially a flip-flop that can be latched in one state or the other.) FFs used in early digital equipment were of the transistor type. In effect, a bistable multivibrator was used. However, gates are used to implement FFs in most present-day digital equipment. The simplest such FF (the basic *reset-set* or RS) is shown in Fig. 2-50 together with the truth table. Note that NAND gates are used and that the gates are cross-coupled (the output of one gate is connected to the input of the opposite gate). The presence of a pulse at either the SET or RESET inputs causes the cross-coupled gates to assume the corresponding state, as shown in the truth table. The gates will remain latched in the state until a pulse is applied at the correct input to change states.

2-13-2   Timing diagrams

A somewhat more advanced circuit, known as a clocked-RS, FF, is shown in Fig. 2-51. This circuit changes states only when both the input pulse and a *clock pulse* are present simultaneously. (The clock pulse is also known as a *gate pulse* or *trigger pulse* in some digital literature.) Operation of the FF in Fig. 2-51 depends on the time relation of pulses as well as the presence of inputs. This relationship can be shown by means of a truth table. However, a timing diagram such as shown in Fig. 2-51 is generally a simpler and more ef- fective way to show time relationships. As an example, in the circuit of Fig.

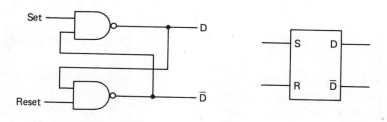

Previous state		Input condition		Result	
D	D̄	Set	Reset	D	D̄
0	1	0	1	1	0
1	0	1	0	0	1
0	1	1	1	No change	
1	0	1	1	No change	
1	0	0	1	No change	
0	1	1	0	No change	
0	1	0	0	1	XXX
1	0	0	0	1	XXX

XXX = unknown.

Figure 2-50 Basic RS FF using cross-coupled NAND gates.

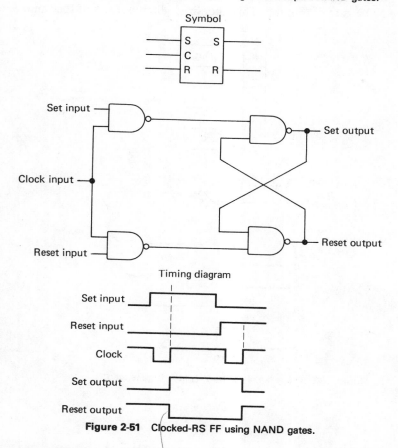

Figure 2-51 Clocked-RS FF using NAND gates.

2-51, the state changes only on the *positive-going edge of the clock pulse* and with the appropriate input present.

Timing diagrams have one major drawback. They do not show what will happen if the inputs are abnormal. The classic example of this occurs when both SET and RESET inputs appear simultaneously. The circuit may move to either state, or remain in its previous state, or may move back and forth between states (known as a *race* condition). Some digital literature includes *flowcharts* or *flow tables* to show the *theoretical* operation of FFs. However, the timing diagram and truth table are generally more practical, and far more realistic.

Obviously, there can be many timing problems produced by the rapid switching between states in a sequential circuit, or even a combinational circuit. The generation of *glitches* is one such problem. Figure 2-52 shows a typical glitch situation where a NOR gate is being fed by an FF that is changing states rapidly. As shown by the truth table, a NOR gate produces a 1 output only when both inputs are at 0. Now assume that the B input changes to 0 slightly faster than the A input changes to a 1. For a time, both inputs are at 0, and the NOR gate produces a 1. Of course, the A input soon becomes a 1, making the NOR gate output a 0. This undesired glitch pulse is of the same

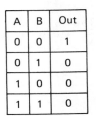

A	B	Out
0	0	1
0	1	0
1	0	0
1	1	0

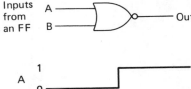

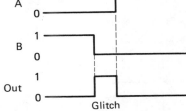

Figure 2-52 One example of how a glitch is generated.

amplitude as other digital pulses on the line (although the glitch pulse is properly shorter in duration). Depending on the circuit, the glitch could cause a serious malfunction or a minor problem such as a flickering of a digital readout. (Note that the term "glitch" is now generally applied to any undesired digital pulse or rapid change of voltage level on a line. In some cases, even noise pulses are called glitches, even though they are produced by external sources and are not necessarily associated with a race condition or any other circuit malfunction.)

2-13-3 Basic FF types

Although there are many variations of FFs in digital circuits, especially those packaged in IC form, there are only three basic types: RS, JK, and master-slave. The basic RS FF is shown in Fig. 2-50. An improved version of the RS FF is the gated RS FF shown in Fig. 2-53. In order to set the gated FF, it is necessary to have two positive pulses appearing simultaneously at the inputs to the SET gate, as shown. The reset condition requires two pulses at the RESET gate.

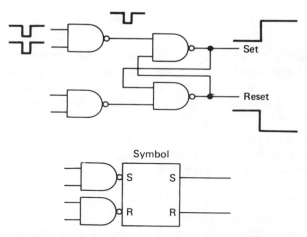

Figure 2-53 Gated-RS FF (with simultaneous pulses at the SET input).

Steered or clocked RS FF. Sometimes one of the pulses used with the SET gate or the RESET gate is a timing or clock pulse, as discussed in Sec. 2-13-2. Often, a single timing or clock pulse can be applied to both the SET and RESET gates. In that case, one input from the SET gate and one from the RESET gate are tied together and appear as a single input, as shown in Fig. 2-54. This type of FF is called a *clocked* or *steered* FF by some digital literature, since the FF requires both a clock pulse and a data input to pro-

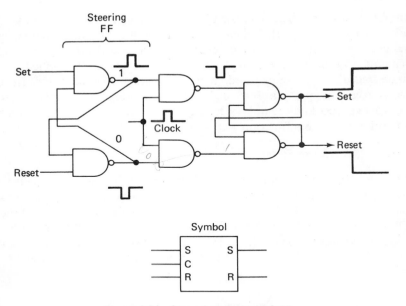

Figure 2-54   Steered or clocked RS FF.

duce a change in output. Other digital literature calls the circuit a steered FF, since the FF must be directed by another element. The element could be a computer data bus, or simply an input from another source. Quite often, the steering element is another FF, the outputs of which are connected to the inputs of the steered FF, as shown in Fig. 2-54.

Self-steered or toggle RS FF.   It is often necessary to change the state of FFs by means of a single trigger (or clock or timing) pulse. This effect can be produced by self-steering, as shown in Fig. 2-55. Such an arrangement is often referred to as the toggle or T-type FF, where the 1 output is connected to the RESET gate input and the 0 output is connected to the SET gate.

This self-steering approach to toggle operation has one major drawback. As soon as the FF changes states, the conditions at the inputs of the SET and RESET gates reverse and, if the timing pulse is still present (this is very possible since the timing pulse duration is usually longer than the delay of the FF), it is quite possible that a second pulse (or glitch) will appear at the output of either the SET gate or the RESET gate. This changes the state of the FF once more. The condition can repeat itself several times (a race condition can occur) until the timing pulse is no longer present. Under these circumstances, it is very difficult to predict the end result. That is, the FF can end up being SET or RESET.

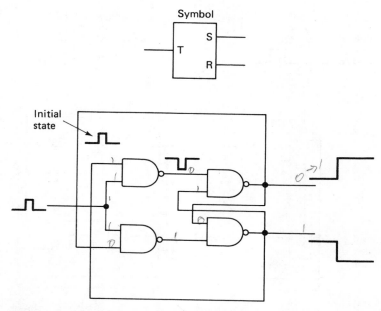

Figure 2-55 Self-steered or toggle (T-type) RS FF.

Cascaded or JK FF. In order to solve the problem of the basic self-steered FF (that is the problem of not knowing the end result of SET or RESET), two gate RS FFs can be cascaded, as shown in Fig. 2-56. The first FF steers the second FF, and the second FF steers the first FF. If FF A is SET, then FF B is SET; if FF B is RESET, then FF A will be RESET.

The common inputs of the two FFs must be separated. This requires two timing or clock pulses. Also, one pulse must be delayed from the other, as shown in Fig. 2-56. For example, if FF B is SET, FF A will be RESET when pulse A arrives. Then if FF A is RESET, FF B will be RESET when pulse B arrives. Since there is no feedback action taking place after the first change for each FF, there are no ambiguous situations, and the end result is predictable.

This approach was used in early digital circuits and is sometimes used (in modified form) today. Early designs of FFs with this capability were known as JK FFs, in contrast to RS FFs. However, the term "JK" is now generally applied to any FF made up of cascaded elements in which the outputs are connected back to the input. Most present-day JF FFs require only one clock input, together with a SET and RESET input, to change states. That is, the FFs can be *preset* to either state. The term "JK" is also applied to the master-slave FF, which operates as follows:

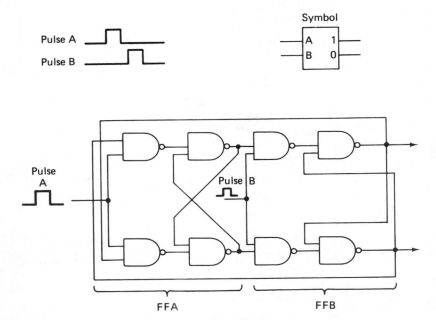

Figure 2-56   Cascaded or JK FF.

Master-slave FF.   The master-slave configuration (shown in Fig. 2-57) is used extensively for FFs in IC form. This circuit is similar to the cascaded FF of Fig. 2-56, with the following two major differences. First, the outputs of the slave FF (Fig. 2-57) are not tied back (internally, within the IC package) to the inputs of the master FF. This is done for greater flexibility. For example, if you want to use only the master or slave portion of the IC, you can do so. In other circuits, both master and slave are used, but you must tie the slave outputs back to the master.

The other major difference is that the common inputs of both master and slave FFs are tied together. This eliminates the need for two clock or timing pulses. With positive logic, the master FF is arranged so that the timing inputs of the SET and RESET gates respond to a positive pulse, while the slave FF responds to negative pulses. The master FF responds to the positive part of a single pulse, and the slave FF responds to the negative part of the same pulse. Typically, the master is triggered by the positive-going edge, and the slave by the negative-going edge of the same pulse.

Note that in the FF circuits discussed thus far, the clock line could be labeled "strobe," "trigger," "timer," "gate," or some similar term. No matter what term is used, when the clock line is zero, the output of the FF cannot be changed by an input at either SET or RESET. Also, the circuits described thus far are *double-rail,* since they provide for two inputs (in addi-

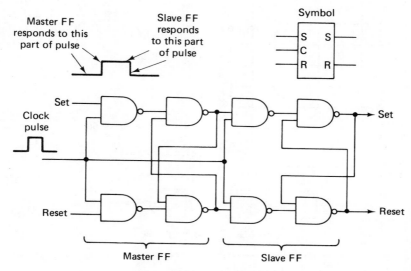

Figure 2-57 Master-slave FF.

tion to the clock input). There are *single-rail* FFs, also called D-type FFs, that require only one data input (in addition to the clock input). Figure 2-58 shows a typical D-type (single-rail) FF circuit together with the truth table. With this circuit, a single data line is connected directly to the SET gate input. An inverter is connected between the data input line and the RESET input. This ensures that the SET and RESET levels cannot be high (or at 1) at the same time.

D-type FFs are especially suited to counter and registers, as described in the following paragraphs. Note that the D-type FF shown in Fig. 2-58 triggers on the leading edge (or positive-going edge, if positive logic is assumed) of the clock pulse. Once the clock line goes to 1, all input changes will affect the FF. Thus, the FF is locked to the input. The data input is not unlocked until the clock input trailing edge (negative-going voltage) falls below the threshold of the input gates.

Double-rank FF. The double-rank FF shown in Fig. 2-59 is similar to the cascaded FF (Fig. 2-56) and the master-slave FF (2-57). The term "double-rank" is used because the FF is essentially two complete toggle-type FFs connected together. The circuit of Fig. 2-59 is often called a JK FF, even though it is not truly JK wired (outputs returned to inputs). The circuit operates on a principle usually referred to as *pulse dodging*. This means that half of the storage cycle occurs on the 0-to-1 transition of the clock line, and the other half occurs on the 1-to-0 transition.

When the clock line is at 0, the output of the master FF is un-

Conditions				New outputs		
New		Old outputs				
Data input	Clock	S	R	S	R	
—	0	1	0	1	0	May not be changed
—	0	0	1	0	1	May not be changed
1	1	—	—	1	0	Set
0	1	—	—	0	1	Reset

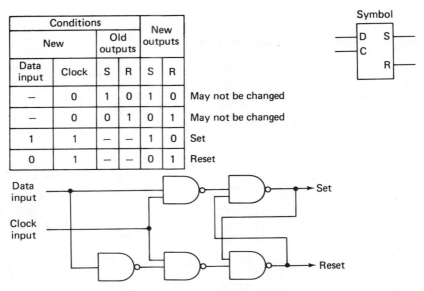

Figure 2-58 D-type (single-rail) FF.

Clock	Inputs		Master	Slave
	R	S		
1	1	1	Same as input	Previous state of slave
1	1	0	Same as input	Previous state of slave
1	0	1	Same as input	Previous state of slave
1	0	0	Previous state of master	Previous state of slave
0	ANY		Previous state of master	Same as master

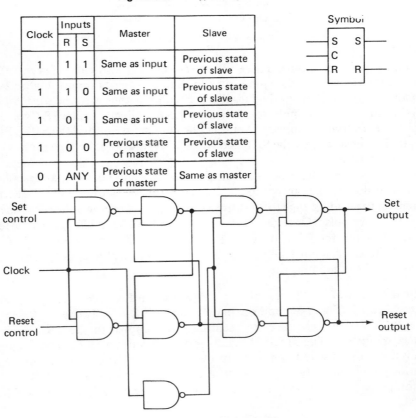

Figure 2-59 Double-rank FF.

changeable, while the output of the slave FF is determined by the state of the master. Thus, when the clock is at 0, the output of slave (and thus the output of the complete FF) is locked to the previous state of the master (last input). On the other hand, when the clock is at 1, the slave FF is locked and the master responds to the input. Such an FF cannot be put into a race condition easily.

Preset FF.   The circuit of Fig. 2-60 is capable of being preset by d-c voltages. Thus, the circuit may be set or reset asynchronously (may be preset, without regard to the clock or any other input) with the PJ and PK inputs. The FF may also be switched synchronously using the J and K inputs together with a clock pulse.

When switched asynchronously, the FF acts like an RS FF (or basic latch), and can be used for *jam transfer* of data: that is, transfer of data into

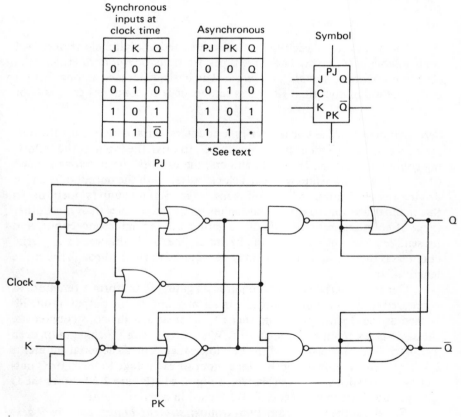

Figure 2-60   JK master-slave FF with preset (PJ and PK).

the FF with no regard for the previous state of the FF. With asynchronous or preset operation, the master and slave are coupled together, and the outputs are set immediately. When switched synchronously, the rising clock pulse cuts the slave off from the master. As the clock line rises still higher, the logic pulses or voltage levels at the J and K inputs are set into the master. Then, when the clock returns to 0, the state of the master is transferred to the slave, which, in turn, sets the output levels.

The synchronous outputs strictly follow the conventional definitions of a JK FF. This is shown on the truth table. If both J and K inputs are at 0, the Q output will remain in the previous state. If both J and K are at 1, the Q output will change states. The preset inputs are also shown on the truth table. Note that the preset (PJ and PK) truth table is identical to the synchronous input (J and K), with one exception. When both PJ and PK inputs are at 1, there is no immediate effect on the outputs. The PJ or PK input that *drops to 0 last* controls the final state of the FF.

2-13-4   Counters

There are three basic types of counters used in digital electronics. All are available in IC form. The three types are *serial or ripple counters, synchronous counters,* and *shift counters.* The basic difference among them is in the method of counting. Counters can be made to count either serially or in parallel.

Serial counters.   Also known as ripple counters, serial counters use the output of a counting element (generally an FF) to drive the input of the following counting elements. In serial counters, the FFs operate in a *toggle* mode (that is, they change state with each clock pulse), with the output of each FF driving the clock input of the following stage. Serial counters operating in this manner are able to operate at higher input frequencies than most other types of counters, require little or no gating and few interconnections, and present only one clock input to the input line. A limitation of all serial counters is that a change of state may be required to ripple through the entire length of the counter.

Generally, serial counters require less gating to perform a function, as compared to synchronous counters. In serial counters, the output of one FF is used as the clock input to another FF, whereas synchronous counter use the same clock input to drive all FFs. With serial, each FF changes on each 1-to-0 (or 0-to-1) transition of the previous stage. The basic serial counter is implemented by connecting the clock input of each stage to the Q or $\bar{Q}$ output of the previous stage, as shown in Fig. 2-61. (Figure 2-60 shows the Q output and clock input of a typical FF used in a serial counter.)

Serial counters fall into two groups: *straight binary* and *feedback.* Straight binary counters divide the input by 2^N, where N is the number of

Basic binary counter

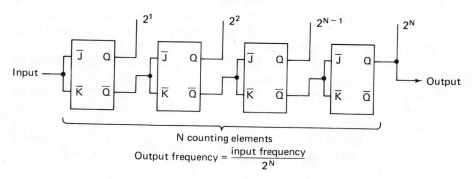

N counting elements

$$\text{Output frequency} = \frac{\text{input frequency}}{2^N}$$

Count
sequence
table

MSB		LSB
C	B	A
0	0	0
0	0	1
0	1	0
0	1	1
1	0	0
1	0	1
1	1	0
1	1	1

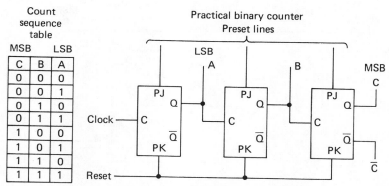

Figure 2-61 Basic and practical binary serial counters.

counting elements. Straight binary counters count in a binary code, with the first counting element (FF) containing the least significant bit (LSB). The basic binary counter of Fig. 2-61 is implemented with JK FFs by connecting the $\overline{J}$ and $\overline{K}$ input together on each FF, and driving this "clock" with the $\overline{Q}$ output of the previous FF. When Q changes from 1 to 0, $\overline{Q}$ changes from 0 to 1 and this changes the state of the following FF. Such a counter will divide by 2, 4, 8, and so on.

A more practical binary counter is also shown in Fig. 2-61. With this counter, the clock input of the LSB FF receives the count line input, and the Q output of each FF is connected to the clock input of the following FF. The count sequence table shows the state of each FF for all counts from 0 to 7. When seven pulses have been applied to the clock line, the A, B, and C FFs will all be at 1. This results in a 111, or binary 7. In this circuit, all FFs can be reset simultaneously. When a 1 is applied to the RESET line (PK inputs), all FFs are reset (the Q output is 0, and the $\overline{Q}$ output is 1). Also, each FF can be preset on an individual basis. This permits a number to be entered in the counter before a count is taken. For example, if it is desired to preset a count of decimal 3 into the counter before the first count, FFs A and B are set to 1, and FF C is set to 0. The first count then changes FFs A and B to 0, and FF C to 1, resulting in 100 or decimal 3.

Feedback serial counters also count in a binary code, but a number of higher-value states are eliminated by the feedback. The basic feedback counter divides the input frequency by $2^{X-1} + 1$, where X is the number of FFs. A division by any arbitrary number can be done using various combinations of straight and feedback counters. A practical feedback counter is shown in Fig. 2-62. The count sequence table for this counter shows the state of each FF for all counts from decimal 0 through 9 in a BCD format.

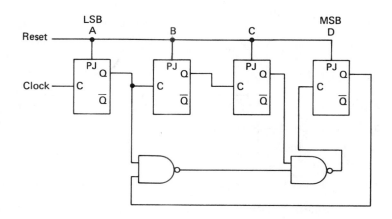

Count sequence table

D	C	B	A	Decimal
0	0	0	0	0
0	0	0	1	1
0	0	1	0	2
0	0	1	1	3
0	1	0	0	4
0	1	0	1	5
0	1	1	0	6
0	1	1	1	7
1	0	0	0	8
1	0	0	1	9

BCD format

Figure 2-62 Feedback ripple counter with BCD format count.

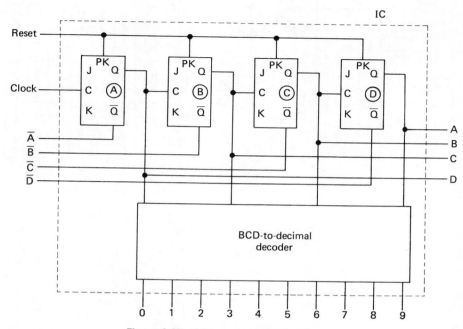

Figure 2-63 BCD counter with decoder.

Many of the counters available in IC form also include auxiliary circuits, such as *decoders, drivers,* and so on. Figure 2-63 is a typical example, in that the circuit is a BCD counter (from 0 to 9), but also provides both a decoded output (indicating the count in decimal form) and the outputs (true and complementary) of each FF stage. Each FF has a RESET function tied to a common RESET line, permitting all stages (FFs) to be RESET (or CLEARED) to 0 simultaneously. The counter is not provided with any preset function. The BCD-to-decimal decoder is similar to that discussed in Sec. 2-12-1 and shown in Fig. 2-41.

Synchronous counters. A synchronous or *clocked* counter is one in which the next state depends on the present state through gating, and all state changes occur simultaneously with a clock pulse. Since all FFs change simultaneously, the output (or count) can be taken from synchronous counters in *parallel* form (as opposed to serial form for ripple counters). Synchronous counters can be designed to count up, count down, or do both, in which case they are often called *bidirectional counters.* The bidirectional (also known as *up-down*) counter is actually an up-counter superimposed on a down-counter. The two counters use the same FFs but have independent gating logic. An up or down signal (usually a fixed d-c voltage) must be pro-

Decimal	Up								Decimal	Down							
	n				n + 1					n				n + 1			
	D	C	B	A	D	C	B	A		D	C	B	A	D	C	B	A
0	0	0	1	1	0	1	0	0	9	1	1	0	0	1	0	1	1
1	0	1	0	0	0	1	0	1	8	1	0	1	1	1	0	1	0
2	0	1	0	1	0	1	1	0	7	1	0	1	0	1	0	0	1
3	0	1	1	0	0	1	1	1	6	1	0	0	1	1	0	0	0
4	0	1	1	1	1	0	0	0	5	1	0	0	0	0	1	1	1
5	1	0	0	0	1	0	0	1	4	0	1	1	1	0	1	1	0
6	1	0	0	1	1	0	1	0	3	0	1	1	0	0	1	0	1
7	1	0	1	0	1	0	1	1	2	0	1	0	1	0	1	0	0
8	1	0	1	1	1	1	0	0	1	0	1	0	0	0	0	1	1
9	1	1	0	0	0	0	1	1	0	0	0	1	1	1	1	0	0

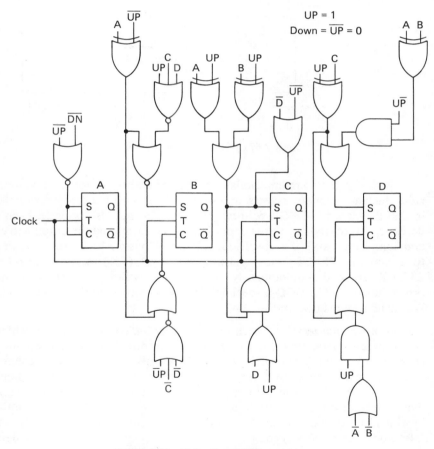

Figure 2-64 Bidirectional XS3 counter.

vided to instruct the counter circuit in which direction to count by activating the proper gating circuitry.

The circuit of Fig. 2-64 is a bidirectional clocked counter using the XS3 code discussed in Chapter 1. As shown in Fig. 1-2, the XS3 (excess 3) code is similar to the BCD code except that the BCD value of three is added to each digit before decoding. The XS3 code is self-complementing, which means that if all the 0s in a number are changed to 1s, and the 1s to 0s, the nine's complement of the number is obtained. This ability to obtain the nine's complement is useful in reducing the required hardware in subtraction operations. Another advantage is that all decimal representations have at least 1 in the XS3 coding. This makes it possible to distinguish zero information from no information.

In the circuit of Fig. 2-64, note that both complementary and true up-down signals are required. Also note that the circuit is shown in simplified form. That is, the gate inputs are labeled as to destination, but the full circuit wiring is not shown. This is common practice for many digital electronic diagrams. The truth table shows the states of each FF during the count, for both up and down counters. The N column represents the present (or existing) stage, while the N + 1 column represents the state at the next bit or clock pulse time.

Shift counters.   Shift counters are a specialized form of clocked counter. The name is derived from the fact that the operation is similar to a shift register (Sec. 2-13-5). In general, the shift counter produces outputs that may easily be decoded, and normally requires no gating between stages (thus permitting high-speed operation). Generally, IC shift counters are found in two forms: *decade shift counters* and *ring shift counters.*

The decade shift counter shown in Fig. 2-65 is sometimes called a switch-tail ring counter, Johnson counter, or simply a shift counter. FFs A through E provide a decade output but require decoding as shown by the output decoding table. For example, if you want a count of 8, you monitor the simultaneous condition of the FF D Q-output and the FF C $\overline{Q}$ output. FF A represents the least significant bit, and FF E is the most significant bit. The output of FF E is fed back into the FF A clock input as shown. This feedback presets the counter to 0 before the counting can begin.

The decade ring shift counter shown in Fig. 2-66 is a typical ring counter with error correction. In some applications it is important to eliminate (or minimize) the decoding logic that follows a counter. A ring counter is useful in such applications. Ring counters use one FF for each count. That is, a 10-count device (or decade counter) requires 10 FFs. Only one FF is true at any time. Thus, no decoding is required.

In the circuit of Fig. 2-66, the K input of the last FF (FF M) is con-

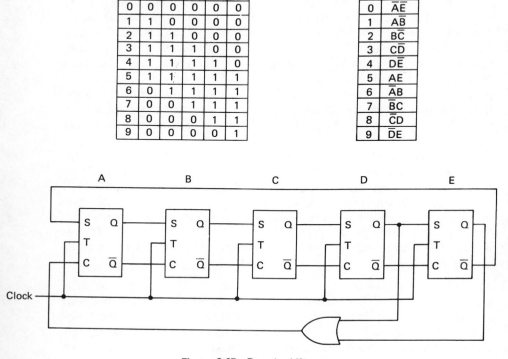

Truth table

	A	B	C	D	E
0	0	0	0	0	0
1	1	0	0	0	0
2	1	1	0	0	0
3	1	1	1	0	0
4	1	1	1	1	0
5	1	1	1	1	1
6	0	1	1	1	1
7	0	0	1	1	1
8	0	0	0	1	1
9	0	0	0	0	1

Output decoding table

0	$\overline{A}\overline{E}$
1	$A\overline{B}$
2	$B\overline{C}$
3	$C\overline{D}$
4	$D\overline{E}$
5	AE
6	$\overline{A}B$
7	$\overline{B}C$
8	$\overline{C}D$
9	$\overline{D}E$

Figure 2-65   Decade shift counter.

nected to a logic 1 level at all times. Note that no output decoding table is provided on Fig. 2-66 (as is the case for a decade shift counter) since a ring counter needs no decoding. For example, in the circuit of Fig. 2-66, if you want a count of 8, you monitor the eighth FF (FF L).

2-13-5   Shift registers and shift elements

The term "register" can be applied to any digital circuit that stores information on a temporary basis. Permanent or long-term storage is generally done with magnetic cores, tapes, disks, drums, and other memories, as discussed in Chapter 3. The term "register" is usually applied to a digital circuit consisting of FFs and gates that can store binary or other coded information.

A *storage register* is an example of such a digital circuit. The various counter circuits discussed in Sec. 2-13-4 are, in effect, a form of storage

Counter sequence

	A	B	C	D	E	F	G	H	L	M
0	1	0	0	0	0	0	0	0	0	0
1	0	1	0	0	0	0	0	0	0	0
2	0	0	1	0	0	0	0	0	0	0
3	0	0	0	1	0	0	0	0	0	0
4	0	0	0	0	1	0	0	0	0	0
5	0	0	0	0	0	1	0	0	0	0
6	0	0	0	0	0	0	1	0	0	0
7	0	0	0	0	0	0	0	1	0	0
8	0	0	0	0	0	0	0	0	1	0
9	0	0	0	0	0	0	0	0	0	1

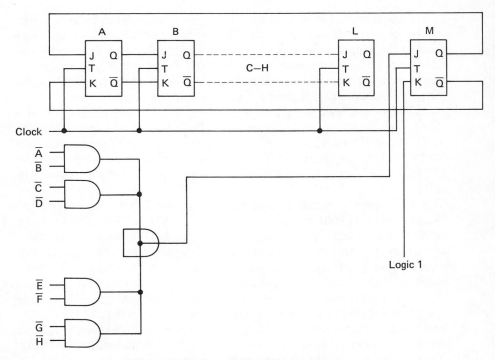

Figure 2-66 Decade ring shift counter.

register. Counters accept information in serial form (the clock or count input) and parallel form (by presetting the FFs to a given count), and hold this information (the count) as long as power is applied and provided that no other information (serial or parallel) is added. With proper gating, the information can be read in or read out.

In many digital circuits, particularly computers, it is necessary to manipulate the data held in registers. For example, registers are used with binary adders and subtractors in arithmetic logic units or ALUs (Chapter 3) to hold and transfer data. Registers can also be used to multiply and divide binary numbers. Both multiplication and division are a form of "shifting." For example, when the binary number 0111 (decimal 7) is shifted one place to the left, it becomes 1110 (decimal 14). Thus, a one-place left shift for a binary number is the same as multiplication by two.

A *shift register* is a circuit for storing and shifting (or manipulating) a number of binary or decimal digits (rather than the simple storage function of a storage register). In addition to their use in arithmetic operation, shift registers are used for such functions as conversion between parallel and serial data.

Storage registers.   Figure 2-67 shows the circuit of a typical binary storage register available in IC form. All FFs are JK type with preset (PJ) and preclear (PK) inputs in addition to the clock or toggle input. Serial information is entered into the register by means of the clock input. Parallel information is entered through the gates and PJ-PK inputs. The circuits are, in effect, ripple counters with gates for parallel entry of data.

The gates are arranged so that the PJ and PK inputs receive opposite states when information is applied to the parallel data lines, and the STROBE line is enabled. For example, assume that the PJ input must be 0 (and the PK input 1) for the Q input to be 1. Under these conditions, the STROBE line is set to 0, and the date line is set to 1, when the Q output is to be 1.

Now assume that a 1 is to be entered into FF A. This presets or "stores" a binary 0001 into the register, since FFA is the LSB (with FF D the MSB). The data input line is then set to 1, with the STROBE line at 0. The 1 appears at the input of gate B, along with a 1 from the STROBE line (inverted from a 0 by gate A). The two 1s produce a 0 output from gate B. This 0 output is applied to the PJ input and sets the Q output of FF A to 1.

At the same time, the 1 input at the D_A line is inverted by gate C and appears as a 0 at the input of gate D, together with a 1 from the STROBE Line. The 1 and 0 inputs produce a 1 output from gate D, which is applied to the PK input. This assures that the Q output of FF A is 1 (with the Q output at 0). All of the remaining FFs receive opposite states: PJ input at 1, and PK at 0. The Q outputs of FFs B, C, and D go to 0, and a binary 0001 results.

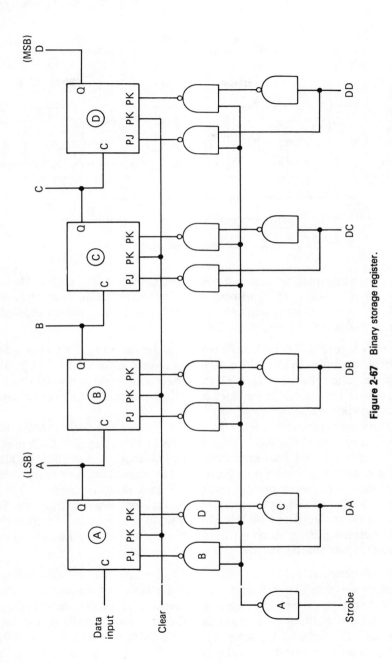

Figure 2-67 Binary storage register.

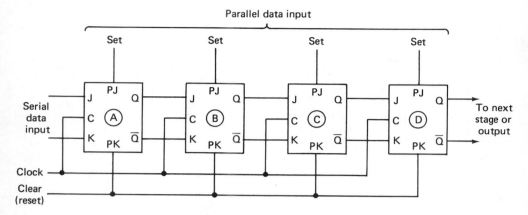

Figure 2-68 Basic shift register.

All the information is cleared when the CLEAR line is enabled. That is, all the Q outputs move to 0 , without regard to their existing state. Data can be taken from the Q or Q̄ outputs, as desired. Serial information is usually taken from the Q or Q̄ outputs of FF D.

Basic shift registers. A basic shift register is shown in Fig. 2-68. (Note that the circuit is similar to the shift counters described in Sec. 2-13.4.) Circuits such as the one shown in Fig. 2-68 shift their contents *one position to the right* for each occurrence of the clock pulse. That is, each stage switches once for each clock pulse.

Data may enter the registers in serial or parallel form from other registers or counters. The RESET line is common to all stages and will preset all stages to 0. All SET lines are independently brought out to allow parallel transfer of data into the register. Both Q and Q̄ outputs of all stages can also be brought out to permit parallel transfer of data out of the register. Thus, data may be shifted serially in and out of the register. When shifting in serial form, the most significant bit (FF D) is transferred out first. To clear the register automatically after all information is shifted out, the J and K inputs of the first stage (FF A) are tied to 0 and 1, respectively.

Multiple function IC registers. Shift registers available in IC form usually have more versatile features than those of the basic register shown in Fig. 2-68. For example, a typical IC register is capable of shifting data stored in the register one position to the right (as with the basic register), or one position to the left, with each clock pulse. When shifting information out of the register, it may be necessary to replace the original information back into the register. In IC registers, this is usually done by an *end-around shift* feature.

For high-speed digital systems, parallel transfer of data is essential. For this reason, IC registers usually provide both serial and parallel transfer, as well as combinations such as serial to parallel, and vice versa. For arithmetic operations, such as subtraction and division (Chapter 3), a *complementation* feature is also desirable. IC registers often include a feature whereby a given set of commands cause the FFs to set up a toggle mode and will complement with the next clock pulse (all FFs with a 0 will go to a 1, and vice versa). There are many means by which to accomplish these functions in IC registers. Unless you are involved in the design of an IC register, the exact means are not critical. However, you should be aware that the features (such as complementing, parallel-to-serial shift, etc.) do exist.

CHAPTER

3

Typical Digital Circuits and ICs

This chapter describes the most common circuits found in a variety of digital electronic systems. All of these circuits are made up of the basic building blocks (gates, counters, and registers) discussed in Chapter 2. Most of these circuits are to be found in IC form. In some cases, an IC will contain one of the circuits. In other cases, several circuits are included on one IC. A typical, present-day digital electronic device is actually a collection of ICs, all interconnected on a common card, board, or module (perhaps with a few miscellaneous resistors, capacitors, transistors, or gates on the same card). In the more sophisticated digital systems involving microprocessors (Chapter 4), all or most of the circuits are combined on a single IC. Before we get into any circuit details, let us start by discussing the various forms of digital ICs.

3-1 DIGITAL IC FORMS

This section describes the various digital IC forms in use today. Some of these forms appeared as discrete component circuits in early digital equipment. However, most of the forms are a result of packaging digital elements as integrated circuits. The digital elements discussed here represent a cross section of the entire digital field.

3-1-1 Resistor-transistor logic (RTL)

RTL was derived from direct-coupled transistor logic (DCTL) and was the first digital IC form introduced, in about 1960. The basic circuit is a direct translation from discrete design into integrated form. RTL was (in the early days of digital electronics) the most familiar to designers, easy to implement, and therefore the first introduced by digital IC manufacturers. The basic RTL gate circuit shown in Fig. 3-1 is presented here to illustrate the basic building block type of logic. The most complex elements are constructed simply by the proper interconnection of this basic circuit. There are many digital devices using RTL still in operation. Thus, RTL is of interest to the student and service technician. However, RTL is not generally used in the design of new digital circuits.

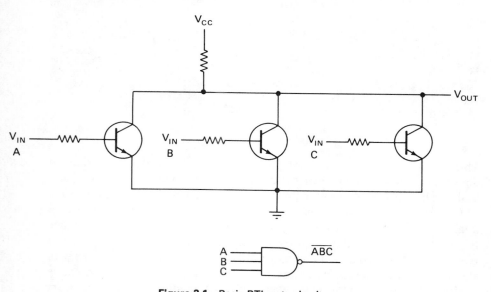

Figure 3-1   Basic RTL gate circuit.

3-1-2 Diode-transistor logic (DTL)

DTL is another digital form that was translated from discrete design into IC elements. DTL was very familiar to the discrete component designer in that the form used diodes and transistors as the main components (plus a

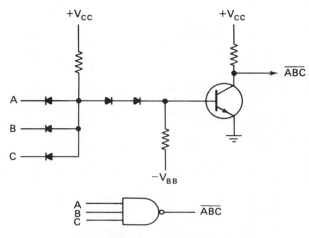

Figure 3-2 Basic DTL gate circuit.

minimum number of resistors). The diodes provide higher thresholds than can be obtained with RTL. Figure 3-2 shows the basic DTL gate. Note that DTL requires two power supplies (to improve the turn-off time of the transistor inverter). As in the case of RTL, DTL is primarily of interest to students and technicians; it is not generally used in the design of new devices.

3-1-3 High-threshold logic (HTL)

HTL is designed specifically for digital systems where electrical noise is a problem but where operating speed is of little importance (since HTL is the slowest of all digital IC families). Because of this slow speed, and because other digital forms (such as the MOS described in Sec. 3-1-6) offer similar noise immunity, HTL is generally not being used in design of present-day digital systems (although HTL is found in many existing systems, particularly where digital circuits must be connected to industrial and other heavy-duty equipment).

Figure 3-3 shows a typical HTL gate (identified as Motorola HTL or MHTL). As shown, HTL is essentially the same as DTL, except that a zener diode is used for D1 in the HTL. The use of a zener for D1 (with a conduction point of about 6 or 7 V), and the higher supply voltage (a V_{CC} of about 15 V), produce wide noise margins. That is, the HTL gate does not operate unless the input voltage swing or pulse is large. Noise voltages below about 5 V have no effect on HTL. The problems of noise in digital circuits are discussed throughout this book.

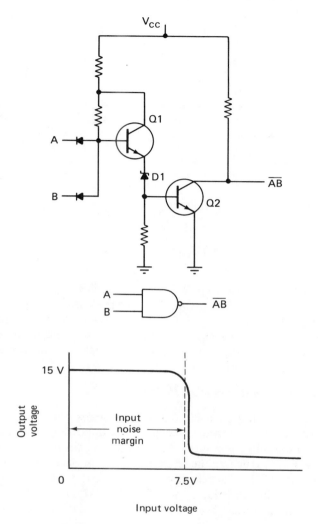

Figure 3-3 Typical HTL gate circuit.

3-1-4 Transistor-transistor logic (TTL or T²L)

ΓTL or (T²L) has become one of the most popular digital logic families available in IC form. Most IC manufacturers produce at least one line of TTL, and often several lines. This fact gives TTL the widest range of digital functions.

Figure 3-4 shows the schematic of a typical logic gate. The circuit shown is that of an IC NAND gate (which is the basic gate for most TTL

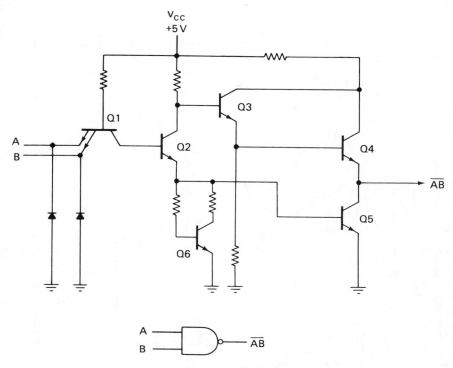

Figure 3-4   Typical TTL NAND gate.

systems). However, a full TTL logic line includes AND, OR, AOI, and NOR, as well as EXCLUSIVE OR and EXCLUSIVE NOR. In addition to basic gates, there is an almost unlimited supply of TTL digital elements, such as FFs, counters, registers, decoders, multiplexers, and line drivers and receivers.

In some digital systems, there are fan-out requirements that exceed the capability of a standard gate. *Power gates* are designed to meet these requirements with a minimum of additional circuitry. A typical power gate (a Motorola MTTL AND gate) is shown in Fig. 3-5. With this power gate, the output circuitry is designed to provide twice the fan-out of conventional gates (in this case, 20 standard gates).

IC line drivers are generally used as amplifiers to increase the fan-out capability of gates, without the use of power gates. Figure 3-6 shows the circuit of NAND line driver together with a typical application of the circuit. Note that the gate output has 75-Ω resistors in series with the standard output (at pins 4 and 5) in addition to the direct output (at pin 6). These resistors provide for terminating the line.

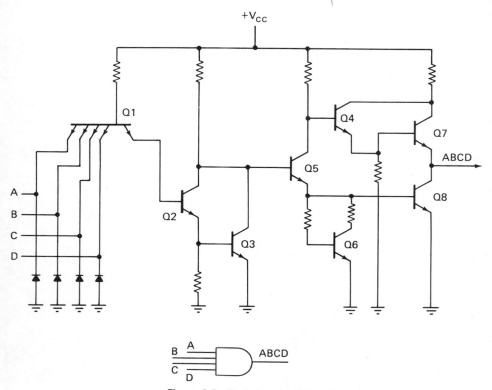

Figure 3-5 Typical power TTL AND gate.

Most TTL gates have an active *pull-up resistor* at the output. This does not permit using a wired-OR operation, as discussed in Sec. 2-8-1. Special gates, such as the MTTL gate shown in Fig. 3-7, are included in many TTL product lines to overcome this limitation. The output of the Fig. 3-7 circuit can be used for wired-OR or to drive discrete components.

Some form of Darlington output is used for most TTL ICs. The Darlington output configuration provides extremely low output impedance in the high state. The low impedance results in excellent noise immunity and allows high-speed operation while driving large capacitive loads. Typically, the high-state output impedance (the state when the logic output is a 1) varies from about 10 Ω at outputs of 3.5 V to about 60 Ω at lower output voltages. The low-state output (logic at 0) output impedance is typically 6 Ω at an output of about 0.2 V, and goes up to about 500 Ω if the output increases to 0.5 V.

One disadvantage of conventional TTL is the so-called "totem-pole" output. As shown in Fig. 3-4, both output transistors are on during a portion

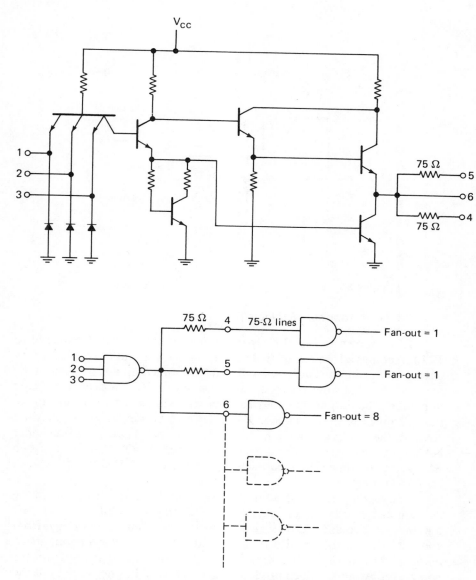

Figure 3-6 Typical TTL NAND line driver with terminating output resistors.

of the switching time. Since the turn-off time of a transistor is normally greater than the turn-on time, a "current spike" can pass through both transistors and the load resistor. The active bypass network (transistor Q6 in Fig. 3-4) helps to limit this problem.

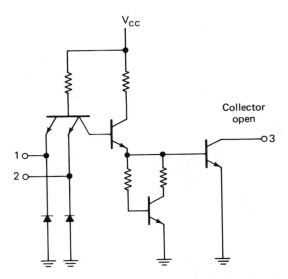

Figure 3-7 Typical TTL gate with open collector for wired-OR use.

3-1-5 Emitter-coupled logic (ECL)

ECL, shown in Fig. 3-8, operates at very high speeds (compared to TTL). Another advantage of ECL is that both a true and a complementary output are produced. Thus, both OR and NOR functions are available at the output. Note that when the NOR functions of two ECL gates are connected in parallel, the outputs are ANDed, thus extending the number of inputs. For example, as shown in Fig. 3-8, when two two-input NOR gates are ANDed, the results are the same as a four-input NOR gate (or a four-input NAND gate connected in negative logic). When the OR functions of two ECL gates are connected in parallel, the outputs are ANDed, resulting in an OR/AND function.

The high operating speed is obtained since ECL uses transitors in the nonsaturating mode. That is, the transistors do not switch full-on or full-off, but swing above and below a given bias voltage. Delay times range from about 2 to 10 nS for a typical gate element. ECL generates a minimum of noise and has considerable noise immunity. However, as a trade-off for the nonsaturating mode (which produces high speed and low noise), ECL is the least efficient. That is, ECL dissipates the most power for the least output voltage.

A typical ECL gate is shown in Fig. 3-9. The tables of Fig. 3-9 illustrate the logic equivalences of the ECL family. As discussed in Chapter 2, it is possible for some digital elements to have two equivalent outputs or functions, depending upon logic definition (positive or negative logic). The ECL

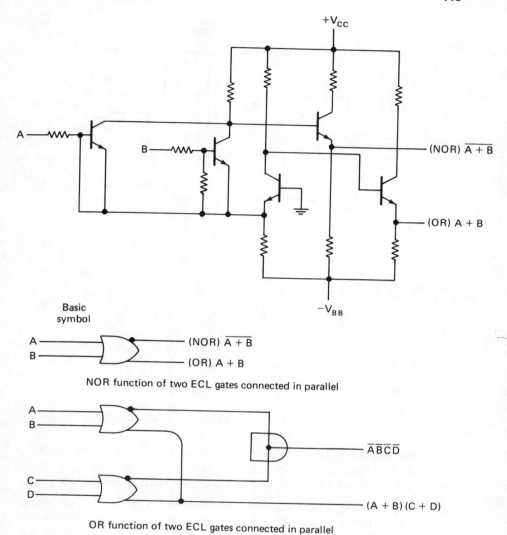

Figure 3-8 Typical ECL gate configurations.

gate shown in Fig. 3-8 can be considered as a NAND in negative logic or a NOR in positive logic. *Saturated logic* families such as TTL have traditionally been designed with the NAND function as the basic logic function. However, in positive logic, the basic ECL function is NOR. Thus, you may design ECL digital systems with positive logic using the NOR, or design with negative logic using the NAND, whichever is more convenient.

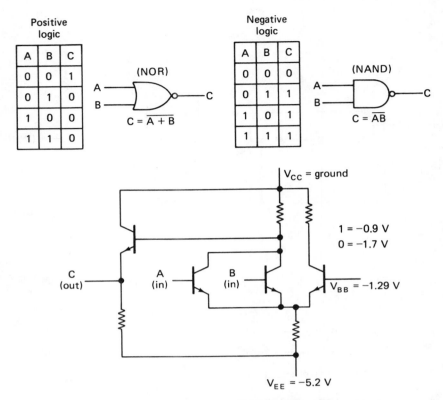

Figure 3-9 Typical ECL gate voltage levels and logic equivalences.

For positive logic, a 1 for the circuit of Fig. 3-9 is about -0.9 V. Logical 0 is about -1.7 V, which makes a nominal voltage switch or pulse of about 0.8 V. (Some ECL lines produce output pulses of about 2 V).

ECL bias. Unlike TTL (and other saturated logic), ECL requires a bias voltage V_{BB}. In the case of the Fig. 3-9 gate, the V_{BB} bias is -1.29 V when the supply voltage V_{EE} is -5.2 V (with V_{CC} at ground or 0 V). With such a gate, if the power-supply voltage is increased (V_{EE} increased due to poor supply regulation, and so on), the 0 level will move more negative, while the 1 level remains essentially constant.

It is essential that the bias voltage V_{BB} track any variations in the power supply voltage. For this reason, some ECL manufacturers provide a *bias driver* with their ECL line. An example of this is the Motorola bias driver down in Fig. 3-10. This bias driver provides a temperature- and voltage-compensated reference for MECL (Motorola ECL). Any of the three MECL voltage may be grounded, but the common voltage of the bias

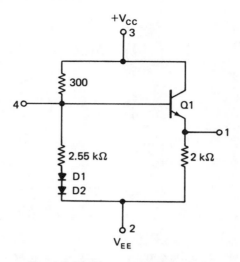

Figure 3-10 Motorola MECL bias driver.

driver must correspond to that of the digital system. If V_{BB} is obtained from the bias driver connected to the *same power supply* as the ECL element, the bias or reference voltage may track the supply voltage changes or temperature variations, thus keeping V_{BB} in the center of the logic levels.

3-1-6 MOS digital ICs

MOS (metal oxide semiconductor) digital ICs are somewhat different from those of other digital ICs, such as TTL, ECL, and so on. MOS ICs are formed using MOSFETs (MOS field-effect transistors) instead of two-junction or bipolar transistors, as do other digital ICs. It is assumed that the reader has a basic understanding of MOSFETs (and other MOS devices). Such subjects are discussed in the author's best-selling *Manual for MOS Users* (Reston Publishing Company, Reston, Va.: 1975), and will not be repeated here. However, before going into details of MOS ICs, let us consider the "what" and "why" of MOS in digital applications.

MOS ICs require only one-third of the process steps needed for two-junction ICs. The most significant feature of MOS ICs is the large number of semiconductor elements that can be put on a small chip. For example, it is possible to put 5000 devices on a MOS chip only 150 × 150 mils square. Each transistor in a typical MOS IC requires as little as 1 square mil of chip area, a great reduction over the typical two-junction IC transistor, which requires about 50 square mils. This high circuit density means *large-scale integration* (LSI), instead of *medium-scale integration* (MSI) or *small-scale integration* (SMI), found in TTL, ECL, and so on.

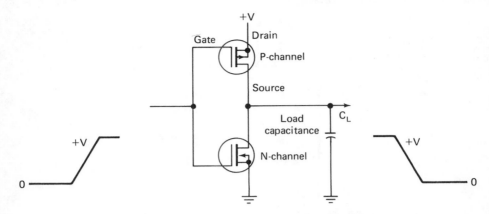

Figure 3-11   Basic MOS device complementary (CMOS) inverter circuit.

Complementary MOS.   Although MOS logic is not limited to the complementary inverter or to the complementary fabrication technique (known as CMOS), the complementary principle forms the backbone of most current MOS digital ICs. The basic complementary inverter circuit is shown in Fig. 3-11. Note that this circuit is formed using an N-channel MOSFET and a P-channel MOSFET on a single semiconductor chip. Some MOS devices are formed using only P-channel (PMOS) or N-channel (NMOS) fabrication techniques.

     The complementary inverter of Fig. 3-11 has the unique advantage of dissipating almost no power in either stable state. Power is dissipated only during the switching interval. And, since MOSFETs are involved, the capacitive input lends itself to direct-coupled circuitry. (The input gate of a MOSFET acts as a capacitor.) No capacitors are required between circuits. This results in a savings in component count and wiring.

     With the input at zero, the upper P-channel MOSFET is fully on. The load capacitance C_L is charged to $+ V$ through the P-channel MOSFET. Once C_L is charged, the only current flow is the I_{DSS} (drain-source current) of the P-channel. Typically, this is in picoampere range, since the MOSFET is completely off with zero gate voltage. The voltage drop across the P-channel MOSFET is the I_{DSS} of the N-channel MOSFET multiplied by the channel resistance of the P-channel. Thus, if the I_{DSS} is 1 pA and the channel resistance is 300 Ω, the voltage drop is about 3 nV. With the input at $+ V$, the N-channel is fully on, and the P-channel is OFF. C_L discharges to ground and the P-channel MOSFET limits current flow to a few pA. Similarly, the voltage drop across the N-channel MOSFET is in the nV region.

Complementary MOS digital NAND gate.   A MOS NAND gate is formed as shown in Fig. 3-12. The P-channel devices are connected in parallel, and

A	B	C	Output
0	0	0	+V
0	0	1	+V
0	1	0	+V
0	1	1	+V
1	0	0	+V
1	0	1	+V
1	1	0	+V
1	1	1	0

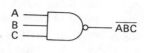

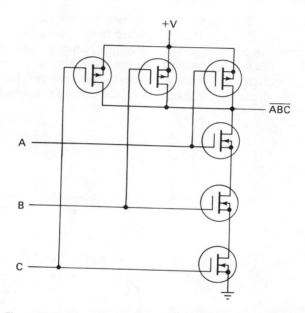

Figure 3-12 Complementary MOS (CMOS) digital NAND gate.

the N-channel complements are connected in series. The truth table for the three-input NAND gate is also given in Fig. 3-12, together with the digital symbol. Note that the output swings from 0 V to +V (which is the supply voltage). If a supply voltage is 10 to 15 V (typical), the difference between a 0 and a 1 is approximately 10 or 15 V. Thus, the MOS digital devices have a much greater noise immunity than the HTL (Sec. 3-1-3). For this reason, and because MOS uses much less power and operates at higher speeds, MOS has generally replaced HTL.

As discussed in Chapter 2, for the NAND function, the output is always 1 unless all three inputs are 1. If any one or any pair of inputs is 1, one or more of the P-channel devices is held on by the remaining 0 inputs, and the output is at $+V$ (1). When all three inputs are 1, all three series N-channels are on, and the output is 0. Note that the 0 output level is developed across the three series elements. However, the leakage current from alll three of the P-channel devices is in the pA range, resulting in nV output voltages when the output is 0. This is an extremely low output voltage, particularly if the normal swing or pulse is in the 10 to 15 V range. As with any solid-state digital device, the limitation on width of the NAND gate (or how long the gate may be held in the 1 or 0 condition) is determined by switching speeds and power dissipation. (Switching speeds decrease, and power dissipation increases, as pulse width increases.)

Transmission gate. The transmission gate is another device unique to MOS digital circuitry. As shown in Fig. 3-13, when the transmission gate is on, a low resistance exists between input and output, allowing current flow in either direction. The voltage on the input line must always be positive with respect to the substrate (V_{SS}) of the N-channel device and negative with respect to the substrate (V_{DD}) of the P-channel device. The gate is on when

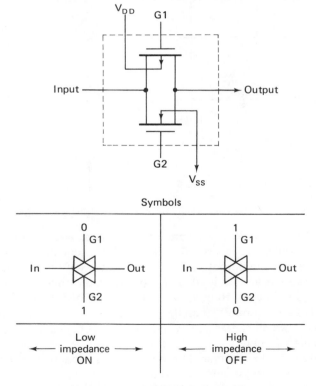

Figure 3-13 Basic MOS transmission gate.

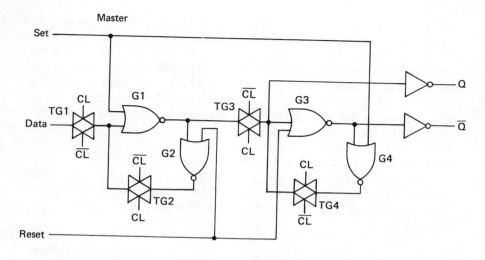

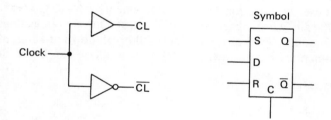

Figure 3-14 MOS IC flip-flop using transmission gates.

the gate G1 of the P-channel is at V_{SS} and the gate G2 of the N-channel is at V_{DD}. When G2 is at V_{SS} and G1 is at V_{DD}, the transmission gate is off, and a resistance of greater than $10^9 \, \Omega$ exists between input and output.

Figure 3-14 shows how the basic transmission gate can be used in a MOS IC flip-flop (a Motorola MC14013). The FF works on the master-slave principle (Chapter 2) and consists of four transmission gates, as well as four NOR gates, two inverters, and a clock buffer/inverter.

3-1-7 Handling and protecting MOS ICs

Damage due to static discharge can be a problem with MOS ICs. Electrostatic discharges can occur when a MOS device is picked up by its case and the handler's body capacitance is discharged to ground through the series capacitance of the device. This requires proper handling, particularly when the MOS IC is out of the circuit. In a digital circuit, a MOS IC is just as rugged as any other solid-state component of similar construction.

MOS ICs are often shipped with the leads all shorted together to prevent damage in shipping and handling (there will be no static discharge between leads). Usually a shorting spring or similar device is used for shipping. The spring *should not* be removed until after the IC is soldered into the circuit. An alternative method for shipping or storing MOS ICs is to apply a conductive foam between the leads. Polystyrene insulating ''snow'' is not recommended for shipment or storage of MOS ICs. Such snow can acquire high static charges which could discharge through the device.

When removing or installing a MOS IC, first turn the power off. If the MOS ic is to be moved, your body should be at the same potential as the unit from which the IC is removed and installed. This can be done by placing one hand on the card or chassis before moving the MOS IC.

Because of the static discharge problem, manufacturers provide some form of protection for a number of their MOS devices. Generally, this protection takes the form of a diode incorporated as part of the substrate material. A diode (or diodes) can be fabricated as part of the chip. The protection scheme used by Motorola in their complementary MOS ICs is shown in Fig. 3-15. The diode is designed to break down at a lower voltage than the MOS device junctions. Typically, the diode will break down at about 30 V, whereas the junctions break down at 100 V. The diode can break down

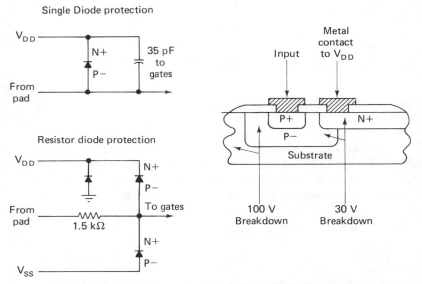

Figure 3-15   Protection methods for MOS static discharge problems.

without damage, provided that the current is kept low (as is usually the case where there is static discharge).

The single-diode scheme provides protection by clamping positive levels to V_{DD}. Negative protection is provided by the 30-V reverse breakdown. The *resistor-diode* protection method adds some delay but provides protection by clamping positive and negative voltages to V_{DD} and V_{SS}, respectively. The resistor is included to provide additional circuit protection.

3-1-8 Comparisons of digital IC forms

At present, there are six basic digital IC families or product lines, three of which (TTL, ECL, and MOS) dominate the field for design of new digital equipment. The following paragraphs summarize the advantages and disadvantages for each of the lines. It is assumed that you will read this material and the discussions showing applications of the various digital IC families throughout this book. You should also study all available datasheets for digital ICs that might suit your requirements.

Availability and compatibility of digital ICs. TTL is the most available of all digital ICs. That is, at present, it is possible to obtain the greatest variety of off-the-shelf digital ICs in the TTL family. Some designers consider TTL as the "universal" IC logic family, since they can obtain an infinite number of gates (with a variety of input/output combinations), buffers, inverters, counters, registers, arithmetic units, and so on, from more than one manufacturer. RTL and DTL were, at one time, the next "most available" digital ICs. Today, both RTL and DTL have been replaced by TTL.

HTL is used only where a high logic swing or pulse (about 13 V) and high noise immunity is required. Except for this advantage, HTL is generally of little value to the designer of modern digital equipment. HTL is the slowest and requires the most power of all digital ICs. In most applications, HTL can be replaced by MOS, since MOS can provide the same logic swing with far less power and at higher speeds. For example, if MOS is operated at a supply power of 15 V, the logic swing can also be almost 15 V.

ECL is used primarily where high speed is essential. ECL is still the fastest of all digital ICs. The disadvantages of ECL are high power consumption (the highest next to HTL) and a low logic swing (usually less than 1 V, but some ECL will provide nearly 2 V).

MOS is the newest of the commonly used digital ICs. The advantages are low power consumption (the lowest of all families), a logic swing equal to any family, and small size. That is, you can get more MOS devices or functions on a given area than any other family. If LSI is required, MOS is the best choice.

TTL and DTL are directly compatible with each other. Thus, some designers use only these two families, taking the advantage of both as applicable; or they design with TTL only when they must adapt new digital circuits to existing DTL systems. RTL is the most compatible with linear and analog systems or any discrete transitor application. This is because RTL is essentially an IC version of conventional solid-state digital circuits.

Because of their special nature, ECL and HTL are the least compatible with other ICs and with external devices. ECL requires a large supply voltage for a comparatively small logic swing, while HTL produces a very large logic swing which is generally too high for other families. MOS can be made compatible with other families, even though the MOS operating principles are quite different.

Keep in mind that, barring some unusual circumstances, any digital IC can be adapted for use with other digital ICs, or with external equipment, by means of *interface circuits* (such as those discussed in Sec. 3-5).

Noise problems in digital ICs.   HTL has the *highest noise immunity* of any digital IC (or HTL has the *least noise sensitivity,* whichever term you prefer). Typically, signal noise up to about 5 V does not affect HTL. Often, HTL can be used without shielding in noise environments where other families require extensive shielding. MOS devices trigger at about 45 to 50 percent of the supply voltage. If MOS is operated at 15 V, the devices will trigger at about 7 V, and will not be affected by lower voltages. Thus, MOS can have higher noise immunity than HTL if the supply voltage can be kept at 15 V.

RTL has the lowest noise immunity. Typically, signal noise in the order of 0.5 V can affect RTL. Considering that RTL operates at a level of about 1 V, an RTL system is almost always operating near the noise threshold. As a result, RTL is not recommended for noisy environments. DTL, TTL, and ECL are about the same with regard to noise sensitivity. However, the input of ECL is essentially a differential amplifier. If one base is connected to a fixed bias voltage about halfway between the 1 and 0 levels, this sets the noise immunity at the level of the bias voltage.

In addition to signal line noise, ICs are affected by *noise on the power supply and ground lines.* Most of this noise can be cured by adequate bypassing, as described in Sec. 3-1-9. However, if there are heavy ground currents due to large power dissipation by the ICs, it may be necessary to use separate ground lines for power supply and logic circuits.

In addition to noise immunity, the *generation of noise by digital ICs must be considered.* Whenever a transistor or diode switches from saturation to cutoff, and vice versa, large current spikes are generated. These spikes appear as noise on the signal lines as well as on the ground and power lines. (As discussed in Chapter 2, such noise spikes are sometimes called glitches.) Since ECL does not operate in the saturation mode, ECL produces the least amount of noise. Thus, ECL is recommended for use where external circuits are sensitive to noise. On the other hand, TTL produces considerable noise and is not recommended in similar situations.

Digital system operating speed.   The speed of a digital IC system is inversely proportional to the propagation delay of the IC elements. That is, ICs with the shortest delay can operate at the highest speed. Since ECL does not saturate, the delay is at a minimum (typically 2 to 4 ns), and speed is maximum. TTL is the next-to-fastest IC family, with delays of about 10 ns, and can be used in any application except where the extreme high speed of ECL is involved. MOS is the slowest of the three currently popular families (TTL, ECL, and MOS). However, MOS is considerably faster than HTL.

Digital IC power requirements.   MOS requires the least power consumption of all digital ICs and can be operated over a wide range of power supply voltages (typically 5, 10, or 15 V). CMOS is well suited for battery-operated systems, since little standby power is required. CMOS uses power when switching from one state to another, but not during standby (except from some power consumed by leakage). TTL and ECL generally operate with a 5-V supply, and consume about 15 and 25 mW (per gate), respectively. The power consumption of a MOS gate is generally figured in the microwatt, rather than the milliwatt, range.

Digital fan-out.   Fan-out, or the number of load circuits that can be driven by an output, is always of concern in digital IC equipment. Some IC datasheets list fan-out as a simple number. For example, a fan-out of 3 means that the IC output will drive three outputs or loads. While this system is simple, it may not be accurate. Usually, the term fan-out implies that the output will be applied to inputs of the same family and the same manufacturer. Other datasheets describe fan-out (or load and drive) in terms of input and output current limits.

Aside from these factors, the following typical fan-outs are available from the digital IC families: RTL 4 to 5, DTL 5 to 8, TTL 5 to 15, HTL 10, ECL 25, and MOS 10.

3-1-9 Practical considerations for digital ICs

It is assumed that you are already familiar with practical considerations applicable to all electronic equipment, particularly IC equipment. For example, you should understand the basics of selecting IC packages, mounting and connecting ICs, and working with ICs (printed circuit, PC boards, lead bending, solder techniques, and so on). It is also assumed that you are familiar with basic electronic test procedures, and that you are capable of interpreting the test information found on digital IC datasheets. If any of these subjects seem unfamiliar, your attention is invited to the author's *Hand-*

All lines AWG No. 20 or larger.
All lines 10 inches (or less)
for each nanosecond of
fastest pulse falltime.

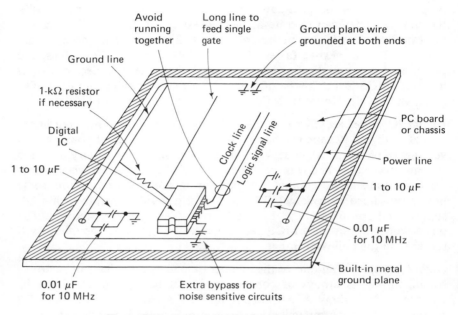

Figure 3-16 Basic digital IC layout considerations.

book of Integrated Circuits (Reston, Va.: Reston Publishing Company, Inc., 1978).

Digital IC layout.   One area that is of particular importance for any digital equipment is the layout of the ICs. The general rules for digital IC layout are shown in Fig. 3-16. The following notes supplement this illustration.

All digital circuits are subject to noise. Any digital circuit, discrete or IC, will produce erroneous results if the noise level is high enough. Thus, it is recommended that noise and grounding problems by considered from the very beginning of layout design.

Whenever d-c distribution lines run an appreciable distance from the supply to a digital module, both lines (positive and negative) should be bypassed to ground with a capacitor, at the point where the wires enter the chassis.

The values for power-line bypass capacitors are typically on the order of 1 to 10μF. If the digital circuits operate at higher speeds (above about 10 MHz), add a 0.01-μF capacitor in parallel with each 1- to 10-μF capacitor.

Keep in mind that even though the system may operate at low speeds, harmonics are generated at higher speeds. The high-frequency signals may produce noise on the power lines and interconnecting wiring. A 0.1-μF capacitor should be able to bypass any harmonics present in most digital systems.

If the digital ICs are particularly sensitive to noise, as is the case with TTL, extra bypass capacitors (in addition to those at the power and ground entry points) can be used effectively. The additional power and ground-line capacitors can be mounted at any convenient point on the board of chassis provided that there is no more than a 7-inch space between any IC and a capacitor (as measured along the power or ground line). Use at least one additional capacitor for each 12 IC packages, and possibly as many as one capacitor for each 6 ICs.

The d-c lines and ground return lines should have large enough cross sections to minimize noise pickup and d-c voltage drop. Unless otherwise recommended by the IC manufacturer, use AWG No. 20 or larger wire for all digital IC power and ground lines.

In general, all leads should be kept as short as possible, both to reduce noise pickup and to minimize the propagation time down the wire. Typically, digital ICs operate at speeds high enough so that the propagation time down a long wire or cable can be comparable to the delay time through a digital element. This propagation time should be kept in mind during the layout design.

Do not exceed 10 inches of line for each nanosecond of fall time for the fastest digital pulses involved. For example, if the clock pulses (usually the fastest in the system) have a fall time of 3 ns, no digital signal line (either printed-circuit or conventional wire) should exceed 30 inches.

The problem of noise can be minimized if ground planes are used (that is, if the circuit board has solid metal sides). Such ground planes surround the active elements on the board with a noise shield. Any digital system that operates at speeds above approximately 30 MHz should have ground planes. If it is not practical to use boards with built-in ground planes, run a wire around the outside edge of the board. Connect both ends of the wire to a common or "equipment" ground.

Do not run any digital signal line near a clock line for more than about 7 inches because of the possibility of *cross-talk* in either direction. If a digital line must be run a long distance, the line should feed a single gate (or other digital element) rather than several gates. External loads to be driven must be kept within the current and voltage limits specified in the IC data sheets. Some digital IC manufacturers specify that a resistor (typically 1 kΩ) be connected between the gate input and the power supply (or ground, depending upon the type of digital IC), where long lines are involved. Always check the IC datasheet for such notes.

3-2 ARITHMETIC LOGIC UNITS (ALU)

This section describes the various ALUs found in digital equipment. ALUs are also called *arithmetic units* or *arithmetic arrays.* Here, we concentrate on explanations of basic principles and techniques used for addition, subtraction, multiplication, and division in digital equipment. A digital circuit that can perform these functions can solve almost any problem, since most mathematical operations are based on these four functions. As discussed in Chapter 4, most microprocessors contain an ALU. Complete ALUs are also available in IC form. In some cases, IC ALUs are interconnected to perform a special function, such as high-speed multiplication. This section also describes such an example.

3-2-1 Adder circuits

Adder circuits are generally used in digital equipment to add binary numbers. There are two basic types of adders: the half-adder and the full-adder.

The *half-adder* is a device for adding two binary bits. The basic circuit, symbol, and truth table for the half-adder are shown in Fig. 3-17. As shown, the half-adder has two inputs, for the 2 bits to be added; and two outputs, one for the sum bit S and the other for the carry-out bit C. The letters HA are often used to denote a half-adder. The half-adder sum is called the EX-CLUSIVE OR function (Chapter 2), or sometimes the "ring sum." The $\oplus$ symbol represents the EXCLUSIVE OR or ring-sum operation. Note that the half-adder of Fig. 3-17 requires complemented inputs. In a typical digital

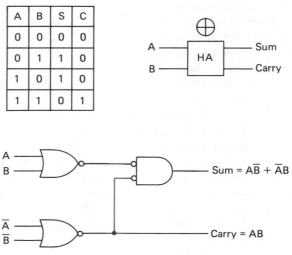

Figure 3-17 Basic half-adder.

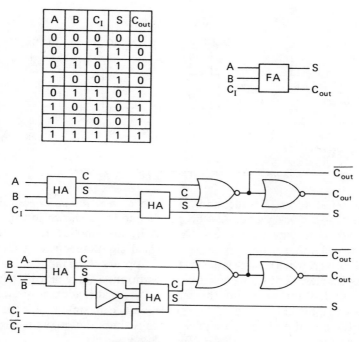

A	B	C_I	S	C_{out}
0	0	0	0	0
0	0	1	1	0
0	1	0	1	0
1	0	0	1	0
0	1	1	0	1
1	0	1	0	1
1	1	0	0	1
1	1	1	1	1

Figure 3-18 Basic full-adder.

adder system, the HA receives its inputs from an FF (or possibly a register) which has complemented outputs.

A *full-adder* is a circuit for adding three binary bits. The circuit, symbol, and truth table for the full-adder are shown in Fig. 3-18. As shown, the full-adder has three inputs (A, B, and C, where A and B represent the two bits to be added and C (or possibly C_{in}) represents the "carry-in" for that stage, which is the "carry-out" from the next-lower-order addition. The two outputs from the full-adder are the sum bit S (which is the sum of the three input bits, A, B, and C_{in}) and the "carry-out" bit C_{out}. The letters FA are often used to denote the full-adder.

Note that there are two full-adder circuits shown in Fig. 3-18. One circuit uses both true and complemented inputs, while the other circuit requires only true inputs. Both circuits are composed of half-adders, gates, and inverters (or gates used as inverters). There are many other circuits for full-adders. However, the truth table for all true full-adders is the same as that shown in Fig. 3-18.

A *parallel adder* adds binary numbers in parallel form (where each bit in a binary number appears on its own line but simultaneously with all other bits in the number). The parallel adder is generally simpler to implement

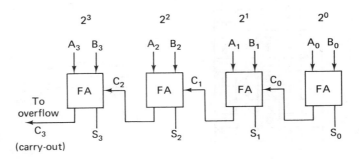

Figure 3-19   Four-bit parallel ripple-carry adder.

than the serial adder. Parallel addition is an *asynchronous* operation, independent of a clock signal. A 4-bit parallel adder is shown in Fig. 3-19. Here, two 4-bit binary numbers (A and B) are applied to the inputs simultaneously. A preliminary sum is formed simultaneously at all positions. However, when a carry is generated, it appears at the input of the next-higher-order state. Thus, in a worst-case condition, a carry bit can "ripple" the full length of the adder. For this reason, the parallel adder is often referred to as a "ripple-carry adder." As a guideline, the maximum time for a ripple to settle out of an N-bit parallel adder is N times the carry propagation delay time per bit.

In a practical parallel adder, the input binary numbers are obtained from a register and are gated into the adder. At a predetermined time later, the sum number S and any carry-out C_{out} are gated out to another register.

A *serial adder* adds binary numbers in serial form (where all bits in a binary number appear on one line, in sequence, usually starting with the least significant bit). Binary serial addition is thus a time-sequential (synchronous) operation, performed one bit at a time. The two least significant bits of both binary numbers to be added (A_0 and B_0) are added first, forming a least significant sum bit (S_0) and a carry bit (C_0). The least significant carry bit (C_0) is then added with the two next-higher-order bits (A_1 and B_1) and the sum and carry bits (S_1 and C_1) are formed; and so on. The two binary numbers to be added are each stored, in binary form, in separate registers. The two numbers are fed serially into an adding network, and the sum bits are fed serially into another register.

The basic serial adder concept is shown in Fig. 3-20(a). Here, the A register contains the augend number and the B register contains the addend. The S register receives the sum number bits serially as they are generated from the full-adder (FA). The carry storage element provides temporary storage for C_{out} until C_{out} is used in the next addition, and the next C_{out} is placed in the storage element. The carry storage element is generally an FF and must be clocked by the same timing pulse that is applied to the registers.

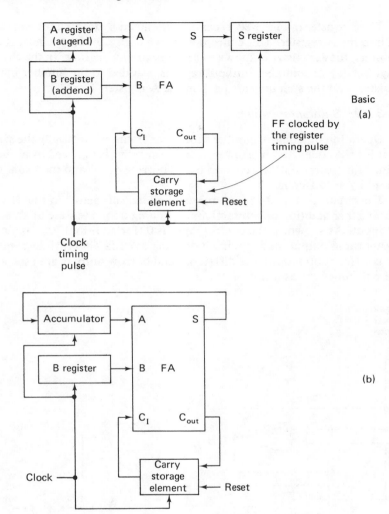

Figure 3-20 Typical serial adder circuits.

After the final 2 bits from A and B are added, the carry storage element contains the final C_{out} which can represent an overflow-alarm bit, if desired. (Register circuits used in arithmetic arrays often include an overflow-alarm network to indicate that the register is full, and can accept no further inputs.) If the A and B registers are large enough that the most significant bit positions in both registers are zeros before the two numbers are added, no overflow condition will occur. The A, B, and S registers operate in a shift-right mode. The augend and addend numbers are loaded into the A and B registers in either serial or parallel mode.

The S register in Fig. 3-20(a) can be eliminated by feeding the sum bits back into the A register. This concept is shown in Fig. 3-20(b). Under these conditions, the A register is known as an *accumulator register.* In the circuit of Fig. 3-20(b), the number in the B register is added to the number in the A-register, and the sum appears back in the A register.

3-2-2   Subtractor circuits

Operation of a binary number subtractor circuit is essentially the same as that for the adder circuit. However, the "carry in" is replaced by a "borrow in," the "carry out" is replaced by a "borrow out," and the "sum" is replaced by a "difference."

The circuit of Fig. 3-21 is a universal adder/subractor, in that it will perform either addition or subtraction, depending upon the state of a "control" input. As shown, if the control input is 0 (for addition), the sum and carry appear as outputs and the circuit operates as a full-adder. If the control input is 1 (for subtraction), the difference and borrow outputs are used and the circuit functions as a full-subtractor.

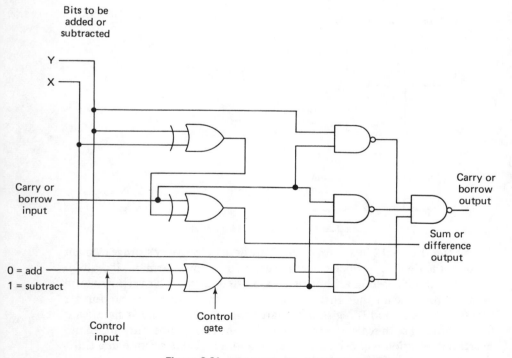

Figure 3-21 Universal adder/subtractor circuit.

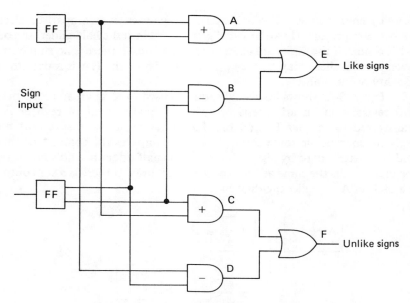

Figure 3-22   Typical sign (polarity) comparator.

3-2-3 Sign and equality comparators

Comparators have many uses in arithmetic units. Typically, a *sign comparator* is used to determine the sign of the numbers (plus or minus) or to compare the polarity of dc voltages. *Equality detectors* compare bits in registers to detect errors. There are many comparator circuits. the following paragraphs describe two typical circuits.

Sign comparator.   As shown in Fig. 3-22, a typical sign comparator consists of two FFs, four AND gates, and two OR gates. The two FFs, each indicating the polarity of an applied voltage, can produce four possible combinations, two for the like polarities and two for the unlike polarities.

If the FFs have the same output, one of the AND gates (A for positive or B for negative) will have an output, and OR gate E provides an output signal. That is, there is an output indicating like polarities whether both FFs have a positive or a negative output. One of the two AND gates provides an output signal, and since an OR gate requires only one input to produce an output, an OR gate E provides a like-polarity output.

If the two polarities are opposites (one positive and one negative), there are only two possible combinations. For either combination, one of the two AND gates (C or D) provides an output to OR gate F, indicating an unlike-polarity output.

Equality comparator.   The half-adder can be used as a comparison circuit for detecting errors. If two unlike digits are fed into the half-adder, the output is a sum. if the digits are alike (0 and 0, 1 and 1), there is no sum output (even though there may be a carry output). Thus, the HA is a detector of equality or inequality.

Figure 3-23 shows how a half-adder is used to compare the contents of two registers in an arithmetic array. The binary count in register A is transferred to register B, and the half-adder must make sure that both registers contain the same count. The HA compares the contents of the A and B registers, digit by digit. As long as the half-adder has no sum output, the digit in A is the same as its respective digit in B. If there is a sum output, the digit in A is unlike the digit in B.

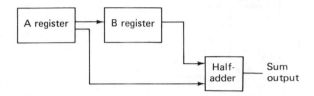

Figure 3-23   Half-adder used as a comparison circuit or equality detector.

3-2-4   Basic ALU and arithmetic array circuits

For this discussion, an arithmetic array is an arithmetic circuit made up of individual ICs (typically adders, subtractors, and registers). An ALU is a single IC that performs the functions of an array (addition, subtraction, multiplication, and division). The following paragraphs describe the basic combinations of circuits that make up either an arithmetic array or an ALU. Section 3-2-5 describes a high-speed multiplication circuit using ALUs in IC form.

Subtraction by complementing.   It is possible to accomplish binary subtraction by complementing (Chapter 1). In a practical circuit, this is done by complementing (or inverting) the subtrahend before it enters either of the inputs (augend or addend) of a full adder. An example of this method is shown in Fig. 3-24, which is a combined parallel adder and subtractor circuit.

When the circuit is to be used for subtraction, the subtract line is made true by a control panel switch or by a programmed pulse. With the subtract line true, the subtract AND gates pass the inverted subtrahend through the OR gate to the corresponding input of the FAs. At this time, the end-around carry gate is opened (by the subtract line being true), inserting C4 (carry four) back into C1 (carry one). The process of complementing and "adding back" the end-around carry produces the correct difference on D1, D2, and

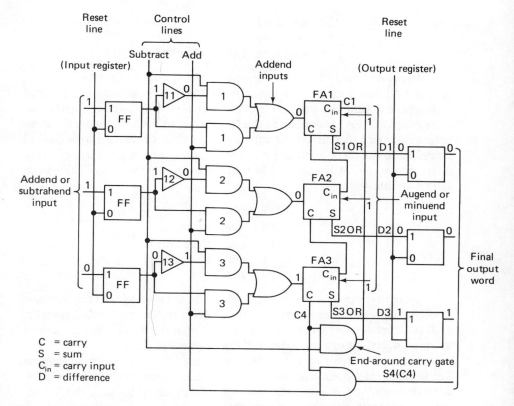

Figure 3-24 Subtraction by complementing using combined parallel adder and subtractor circuits.

D3. For example, assume that 011 is to be subtracted from 111. Using normal binary subtraction,

$$\begin{array}{r} 111 \\ -011 \\ \hline 100 \end{array} \quad \begin{array}{l} \text{minuend} \\ \text{subtrahend} \\ \text{difference} \end{array}$$

Now, using the complement method circuit of Fig. 3-24, the 111 minuend is entered into the minuend input of FA1, FA2, and FA3. The 011 subtrahend is entered at the inverters so that 0 is applied to I3, 1 is applied to I2, and the least significant 1 is applied to I1. The inverters produce the complement 100, which appears at the subtract AND gates (1 at gate 3, 0 at gate 2, and least significant 0 at gate 1). In turn, the normal addend input of all three adders receive this complement. This results in the addition of 111 and 100, or

```
 111    minuend
+100    complement
 011    sum before and end-round carry
```

However, the end-around carry (C4) of 1 is applied to the carry input (C1) of FA1. Since the sum-before carry of FA1 is 1, the end-around carry of 1 changes the sum (now the difference) output of FA1 to 0. Similarly, the carry output of FA1 is applied to the carry input of FA2, changing the sum-before carry of 1 to a difference output of 0. In turn, the carry output of FA2 is applied to the carry input of FA3, changing the sum before carry of 0 to a difference output of 1, with no addition carry output from FA3. This can be shown as follows:

```
         111    minuend
        +100    complement
   1     001    sum-before carry
        +  1    end-around carry
         100    difference
```

Note that the registers shown in Fig. 3-24 are used to hold the numbers or count at the input and output of the adder/subtractor circuit. Typically, such registers are made up of FFs that can be reset to zero (by the operator or a timing pulse) before and after arithmetic operations.

When the circuit of Fig. 3-24 is to be used for addition, the addition line is made true, permitting the add AND gates to pass the noninverted addend through the OR gate to the corresponding addend input of the full adders. The circuit then acts as a conventional adder.

Multiplication by shifting. It is possible to accomplish binary multiplication by adding and shifting, because multiplication is the process of adding the multiplicand to itself as many times as the multiplier dictates. For example:

```
     1111    multiplicand
   ×1101    multiplier
     1111    A ⎫
     0000    B ⎬  partial products
     1111    C ⎪
     1111    D ⎭
 11000011    product
```

In binary multiplication (where only 1s or 0s are used), the partial produces are always equal to zero or to the multiplicand. The final product is obtained by the addition of as many partial products that are equal to the multiplicand as there are 1s in the multiplier. In our example, there are three 1s in the multiplier. Thus, three partial products (A, C, and D) are equal to the multiplicand and must be added. Of course, the addition must take place in the proper order, since each partial product is shifted by one digit for each digit in the multiplier (where there is a 1 or 0). In our example, there are four digits in the multiplier. Thus, there must be four shifts.

Figure 3-25 shows a typical shift multiplier circuit. Note that the circuit consists of three registers (multiplicand, multiplier, and accumulator) all made up of FFs, gates, and inverters; three parallel FAs; and a shift FF. The shift FF provides two bit times for each step of operation. Thus, one bit time is required for adding or not adding, and a second bit time is necessary for the shift. The shift FF alternately sets and resets, thus providing a true signal (S) to the various gates every other bit time. The output of the shift FF controls the action of the "add" or "not add" during one bit time (S*) and the shift during the next bit time (S).

Assume that the circuit of Fig. 3-25 is to multiply binary 101 by 111 (5 × 7 = 35). All the multiplier and multiplicand register FFs are reset to zero; then the appropriate digits (1 or 0) are set into each FF as applicable. (The reset and preset lines are omitted from Fig. 3-25 for clarity.)

The multiplication process is started by a clock pulse, which resets the shift FF, making the S output false and the S* output true. Thus, the first action that can occur (in conjunction with the first clock pulse) is an add or a not-add, depending on the state of multiplier FF FMR3.

With the shift FF reset, the S* output is true. The output of FMR3 is also true, since the multiplier register is initially set to 101. Under these conditions, all the parallel adder AND gates are enabled. The contents of the multiplicand register (FMC1–FMC3) and the accumulator register (FA2–FA4) are added. Then the sum and carry outputs of the full adders are put back into the accumulator register FFs (FA1–FA4). Since the accumulator register is initially reset to all 0s, the sum is equal to the multiplicand (111), and this value is the sum answer in the accumulator FFs FA2–FA4.

At the next clock pulse, the S output is true and the S* output is false. Thus, no more adding can take place, but a shift occurs in both the multiplier and accumulator registers (since the AND gates for these registers are enabled by the S output).

At the next clock pulse, the S* is again true and the S is false. The full-adders again try to add the contents of the multiplicand register (FMC1–FMC3) to the accumulator register (FA2–FA4). However, the output of multiplier FF FMR3 is now false, keeping the full-adder gates closed. FF FMR3 is false because it has been shifted one digit by the clock pulse. Originally, FMR1 was at 1, FMR2 at 0, and FMR3 at 1. After the shift, however, both FMR1 and FMR3 are at 0 (false), with FMR2 at 1.

$$\begin{array}{r} 111 \\ 101 \\ \hline 111 \\ 11 \end{array}$$

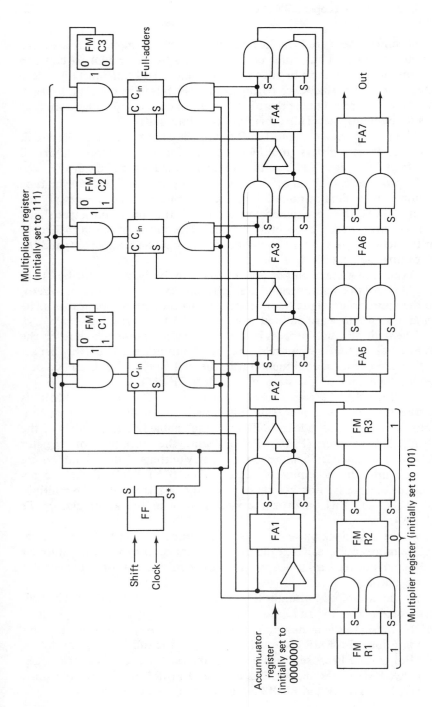

Figure 3-25 Typical shift multiplier circuit.

At the next clock pulse (S* false, S true), another shift occurs. The accumulator register FFs are now at FA1 = 0, FA2 = 0, FA3 = 0, FA4 = 1, FA5 = 1, FA6 = 1, and FA7 = 0 (0001110).

At the next clock pulse (S* true, S false), an add does take place, since the MSD of the multiplier (101) has shifted into FMR3. The addition that takes place is:

$$\begin{array}{ll} 111 & \text{multiplicand register} \\ \underline{0001110} & \text{accumulator register} \\ 100011 & \text{final true product} \end{array}$$

Although the correct product now exists in the accumulator register, the next shift takes place automatically, placing the product's LSD in FA7.

In a typical arithmetic array or ALU (such as one found in a microprocessor, Chapter 4), a *stop multiplication pulse* from the control section halts action of the multiplication circuit at this time. In some ALUs, the contents of the accumulator register can be read out by front-panel lights (sometimes called *register lights*) connected to show the state of the individual accumulator FFs (the light is on if the FF is in the 1 state). Typically, the next operation in the computer program is to transfer the contents of the accumulator register (now 0100011, or decimal 35) to another register, or perhaps into the computer memory (Sec. 3-6), for further processing.

Simultaneous multiplication.   In some ALUs, particularly those used in special-purpose computers, operations must be performed at high speeds. Instead of adding (or subtracting), complementing, and shifting two sets of binary digits, each digit is processed simultaneously with all other digits. A separate network (adder, subractor, control gates, and so on) is required for each digit, similar to the circuit of Fig. 3-24. Of course, the circuits that provide simultaneous arithmetic operations add many components to the array (thus increasing the size and cost).

Figure 3-26 shows a circuit for simultaneous multiplication of two four-digit binary numbers, producing an 8-bit output. The best way to understand operation of this circuit is to trace through the multiplication of two numbers. The binary numbers used in the explanation of Fig. 3-25 (111 × 101) are again used in Fig. 3-26. The input and output states of all gates and adders are shown. Note that the final product is 00100011 (decimal 35).

Repeated-addition multiplication.   Since multiplication is the process of adding the multiplicand to itself as many times as the multiplier dictates, it is possible to perform multiplication by repeated addition. For example, to multiply 111 or 101 (7 × 5), the multiplicand 111 is added to itself 101 (or 5) times. Some ALUs use the repeated-addition system of multiplication. However, the trend today is toward the shifting method (for general-purpose use) or simultaneous multiplication (for high-speed use).

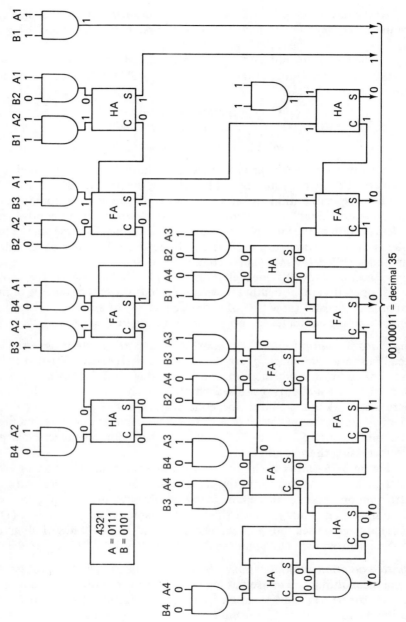

Figure 3-26 Simultaneous multiplication circuit.

00100011 = decimal 35

4321
A = 0111
B = 0101

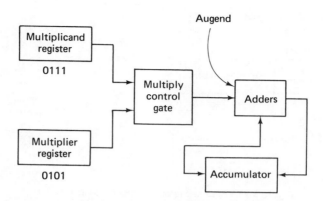

Figure 3-27 Basic repeated-addition multiplication circuit.

Figure 3-27 is a block diagram of a repeated-addition multiplier circuit where the value of the multiplicand is put into a multiplicand register, and the value of the multiplier is put into a multiplier counter. The counter counts down (from the multiplier value) to 0. For example, if the multiplier is 101, the counter registers 101 before the first count, 100 after the first count, and 011 after the second count, and so on, down to 000. At count zero, the multiply control gate is closed, halting the repeated-addition process.

As long as the gate is open, the value of the multiplicand register is added to the existing value in the accumulator register. Note that the accumulator must have as many FFs as the sum of the multiplier and multiplicand. In our example (101 × 111) there is a total of six digits. Thus, the accumulator must have at least six digits. The multiplicand and accumulator outputs make up the augend and addend inputs to the adder circuit. The accumulator starts at 0 and holds the answer from the last addition. Thus, the accumulator value increases as the repeated addition continues.

Division circuits. As with multiplication, a separate circuit is sometimes used in ALUs to perform division only. Such circuits are used where arithmetic operation must be performed at high speeds, without regard to increased size and cost. High-speed circuits for simultaneous division are similar to those for multiplication (Fig. 3-26) except that subtractors and gates are used (instead of adders and gates) since division is essentially a process of repeated subtraction, or subtraction and shifting.

The *basic repeated-subtraction divider* circuit is shown in the block diagram of Fig. 3-28, where the value of the divider is put into a division register, and the value of the dividend is put into the accumulator. The divisor and accumulator output make up the subtrahend and minuend inputs to the subtractor circuit. The counter starts at zero and holds the partial quotients until the last subtraction takes place.

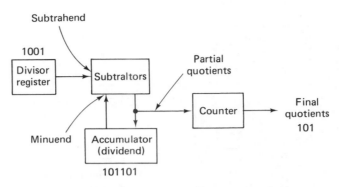

Figure 3-28 Basic repeated-subtraction divider circuit.

As each subtraction occurs, the value in the accumulator becomes smaller, while the count in the counter become larger, finally producing the correct quotient. One problem with such a divider is that it does not stop when the accumulator reaches 0 (or some remainder). Instead, the subtractor tries to continue subtracting beyond the point of the 0 remainder. When this occurs, the value in the accumulator register becomes a negative quantity. In a practical circuit, this problem is overcome with a sign comparator (Sec. 3-2-3). When the sign comparator notes a negative quantity in the accumulator, the divide process is stopped and the overage is added back to the accumulator total.

The basic *subtraction-shifting divider* is shown in the block diagram of Fig. 3-29, where the dividend is put into the accumulator and the divisor is put into the divisor register. The quotient register holds the quotients produced by repeated subtraction.

For each step (or timing pulse) a shift occurs in the accumulator and quotient register. Thus, a 0 or a 1 is posted in the quotient register for each step. A 0 is posted when a subtraction cannot be performed, and a 1 is posted when a subtraction can be performed. Note that the divisor is fed into the subtractor in a manner corresponding to the digits of the highest order of the dividend. The 1s and 0s posted in the quotient register become the final quotient, as shown by the sample division problem of Fig. 3-29.

3-2-5 High-speed binary multiplication

This section describes operation of a high-speed binary multiplication system using Motorola MECL ALUs. Fast arithmetic processing often requires high-speed multiplication of binary numbers. This high-speed multiplication has usually been implemented with a ripple processor using various techniques (requiring large numbers of interconnects and parts). Parallel multiplication, on the other hand, requires large numbers of high-speed adders.

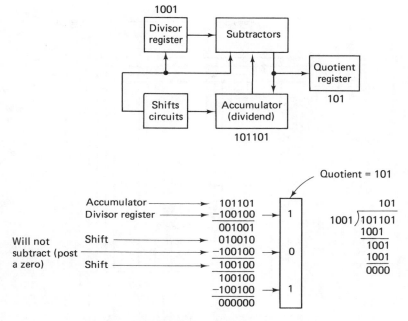

Figure 3-29 Basic subtraction shifting circuit.

By using the fast add time (typically a 4-bit add in 7 ns) and the versatility of an ALU, package count can be significantly reduced. Figure 3-30 shows the diagram of a 4-bit-by-4-bit multiplier using a simple add-shift technique. Each level of logic performs an add function, followed by a hard-wired shift. The equivalent numerical tabulation for this operation is also shown in Fig. 3-30. The first add operation is performed by a gate package because no carry can result at that level. Each succeeding level is performed by an ALU, and forms the next partial product, until the final product is reached.

Motorola MC10181 ALU. To understand operation of the multiplier circuit of Fig. 3-30, it is necessary to understand the operation of the ALU. Figure 3-31 shows the symbol and logic function table for the ALU. As shown, the MC1018 is capable of performing 16 logic operations, and 16 arithmetic operations, on two 4-bit operands. The MC10181 has inputs for a 4-bit operand A, a 4-bit operand B, and a carry-in. The outputs provide for a 4-bit function-out (F0-F3), carry-out ($C_N + 4$), group carry propagate (P_G), and carry generate (G_G). Not all of these outputs are used in the high-speed multiplier.

The choice of logic or arithmetic operation of the ALU is determined by internal circuits controlled by the mode (M) input. In arithmetic mode,

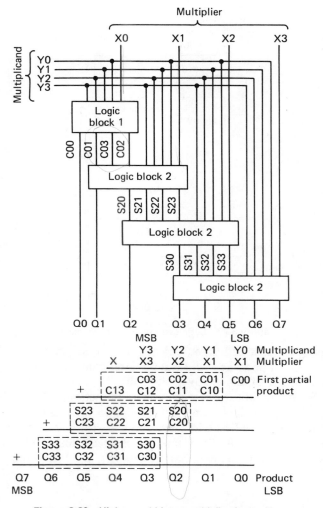

Figure 3-30 High-speed binary multiplication system.

full internal lookahead carry is incorporated to determine the 4-bit function-out and carry-out. The mode control disables the internal carry circuitry for logic mode operation, and the carry is ignored. In the logic mode, the function-out is generated on a bit-by-bit basis.

Each mode of operation has a 16-function capability (32 total). The four select lines (S0-S3) determine which function is performed on operands A and B in either mode of operation.

In logic mode, the ALU can perform any of 16 possible logic opera-

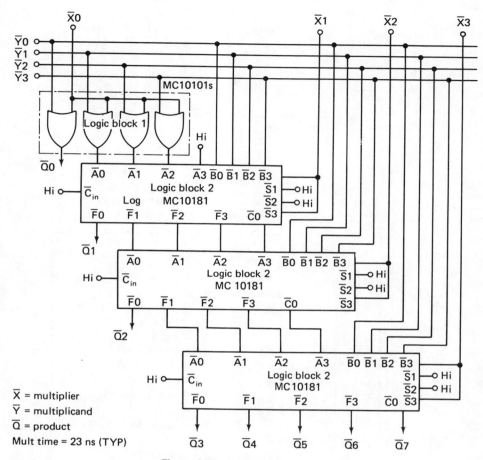

Figure 3-30 (*continued*)

tions on two variables. These functions are especially useful for applications (such as computer processors) where the processor must execute commands, such as generation of all 1s or all 0s, "mask" the contents of a register, OR or AND the contents of two registers, or perform similar logic operations.

The arithmetic mode is also composed of 16 functions. These functions perform various forms of addition and subtraction on operands A and B (which, in turn, can be used for multiplication and division). The different functions are useful for a variety of operations. As examples, the A times 2 function may be used for an arithmetic shift, the A minus 1 function does a decrement (countdown) operation, the two's complement (or F minus 1) function may be generated, and the A plus 0 function may either ripple through word A, or increment word A, by adding a carry-in.

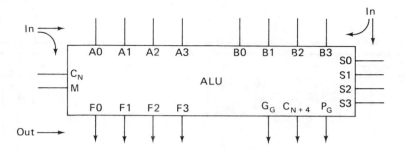

Positive logic

Function selection				Logic functions M is high C = D.C. F	Arithmetic operation M is low C_n is low F
$\overline{S3}$	$\overline{S2}$	$\overline{S1}$	$\overline{S0}$		
L	L	L	L	$F = \overline{A}$	$F = A$
L	L	L	H	$F = \overline{A} + \overline{B}$	$F = A$ plus $(A \cdot \overline{B})$
L	L	H	L	$F = \overline{A} + B$	$F = A$ plus $(A \cdot B)$
L	L	H	H	$F = $ logical "1"	$F = A$ times 2
L	H	L	L	$F = \overline{A} \cdot \overline{B}$	$F = (A + B)$ plus 0
L	H	L	H	$F = \overline{B}$	$F = (A + B)$ plus $(A \cdot \overline{B})$
L	H	H	L	$F = A \odot B$	$F = A$ plus B
L	H	H	H	$F = A + \overline{B}$	$F = A$ plus $(A + B)$
H	L	L	L	$F = \overline{A} \cdot B$	$F = (A + \overline{B})$ plus 0
H	L	L	H	$F = A \oplus B$	$F = A$ minus B minus 1
H	L	H	L	$F = B$	$F = (A + \overline{B})$ plus $(A \cdot B)$
H	L	H	H	$F = A + B$	$F = A$ plus $(A + \overline{B})$
H	H	L	L	$F = $ logical "0"	$F = $ minus 1 (two's-complement)
H	H	L	H	$F = A \cdot \overline{B}$	$F = (A \cdot \overline{B})$ minus 1
H	H	H	L	$F = A \cdot B$	$F = (A \cdot B)$ minus 1
H	H	H	H	$F = A$	$F = A$ minus 1

Figure 3-31 Symbol and logic function table for ALU.

Among other functions, the MC10181 has the ability to sum (word A) + (word B), or to sum (word A) + (ZERO), depending on the state of the control inputs (SO–S3). This feature is to "mask" the multiplicand (word B) and add (ZERO) to the previous partial product of the multiplier, and then add, to form the next partial product.

Note that in the circuit of Fig. 3-30, the one's complement of the multiplicand and multiplier are used to provide the one's complement of the product. This is necessary because of the basic NOR function definition of the MC10181. The typical multiply time of the Fig. 3-30 circuit can be calculated from the worst-case delay path through the multiplier. Multiply time is equal to one gate propagation delay, plus three times the "word-A-to-function-out" delay of the ALU. The delay is typically 2 nS + (3 × 7) = 23 nS.

The multiplier may be expanded to accommodate larger numbers of bits. To multiply N bits by N bits, each adder must be N bits wide. Also, there are N logic levels, with the first logic level implemented with gates, and N − 1 secondary logic levels implemented using ALUs.

3-3   DIGITAL-TO-ANALOG AND ANALOG-TO-DIGITAL CONVERSION

This section describes the various digital-to-analog (D/A) and analog-to-digital (A/D) techniques found in digital electronic equipment. Here, we concentrate on explanations of the basic principles and techniques of D/A and A/D converters. By studying this information, you should be able to understand operation of any D/A or A/D system. The section also includes a specific example of state-of-the-art D/A and A/D conversion.

3-3-1   Conversion between analog and digital information

There are several methods for converting voltage (or current) into digital form. One method is to convert the voltage into a frequency or series of pulses. Then the pulses are converted into BCD form and the BCD information is converted into a decade readout. This is the system most often used in digital meters and frequency counters.

It is also possible to convert voltage directly into BCD form by means of a D/A converter. Before discussing the operation of these conversion circuits, let us discuss the signal formats for BCD data, as well as the 4-bit system.

Typical BCD signal formats.   Although there are many ways in which pulses can be used to represent the 1 and 0 digits of a BCD code, there are three ways in common use. These are the NRZL (nonreturn-to-zero-level), the NRZM (nonreturn-to-zero-mark), and the RZ (return-to-zero) formats. Figure 3-32 shows the relation of the three formats.

In NRZL, a 1 is one signal level and a 0 is another signal level. These levels can be 5 V, 10 V, 0 V, − 5 V, or any other selected values, provided that the 1 and 0 levels are entirely different.

In RZ, a 1 bit is represented by a pulse of some definite width (usually a one-half bit width) that returns to zero signal level, while the 0 bit is represented by a zero-level signal.

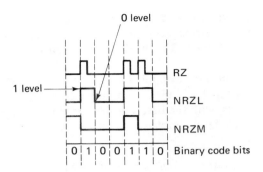

Figure 3-32 Typical BCD signal formats.

In NRZM, the level of the pulse has no meaning. A 1 is represented by a change in level, and a 0 is represented by no change.

The 4-bit system in D/A and A/D conversion.   As discussed in Chapter 1, a 4-bit system is one that is capable of handling 4 information bits. Any fractional part (in sixteenths) can be stated by using only 4 bits. The number 15 is represented by 1111 and zero is represented by 0000. Any number between 0 and 15 requires only 4 bits. For example, the number 1000 is 8, the number 0111 is 7, and so on. Although not all digital codes use the 4-bit system, it is common and does provide a high degree of accuracy for conversion between analog and digital data. Most conversion systems use the 4-bit code or a multiple of the basic code (8 bits, 16 bits, and so on).

In practice, a 4-bit A/D converter (also called a *binary* encoder in some digital literature) samples the voltage level to be converted and compares the voltage to ½ scale, ¼ scale, ⅛ scale, and ¹⁄₁₆ scale (in that order) of some given full-scale voltage. The A/D converter (or encoder) then produces 4 bits, in sequence, with the decision (or comparison) made on the most significant (or ½ scale) first.

Figure 3-33 shows the relation among three voltage levels to be converted and the corresponding binary code (in NRZL form). As shown, each of the three voltage levels is divided into four equal time increments. The first time increment is used to represent the ¹/₂-scale bit, the second increment is tuned for the ¹/₄-scale bit, the third increment for the ¹/₈-scale bit, and the fourth increment for the ¹/₁₆-scale bit.

In level 1, the first two time increments are at binary 1, with the second two increments at 0. This produces a 1100, or decimal 12. Twelve is three-fourths of 16. Thus, level 1 is 75 percent of full scale. For example, if full scale is 100 V, level 1 is at 75 V.

In level 2, the first two increments are at 0, while the second two increments are at 1. This is represented as 0011, or 3. Thus, level 2 is three-sixteenths of full scale (or 18.75 V).

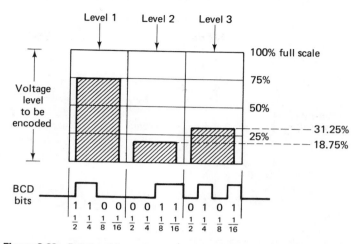

Figure 3-33 Relationship among three voltage levels to be encoded and the corresponding BCD code (using the 4-bit system).

This can be expressed in another way. In the first or half-scale increment, the converter produces a 0 because the voltage (18.85) is *less than* half-scale (50). The same is true of the second, or one-fourth scale increment (18.75 V is less than 25 V). In the third or one-eighth scale increment, the converter produces a 1, as it does in the fourth or one-sixteenth scale increment, because the voltage being compared is *greater than* one-eighth of full scale (18.75 is greater than 12.5) and greater than one-sixteenth of full scale (18.75 is greater than 6.25). Thus, the one-half and one-fourth scale increments are at 0, while the one-eighth and one-sixteenth scale increments are at 1. Also, $\frac{1}{8} + \frac{1}{16} = \frac{3}{16}$, or 18.75 percent.

A/D conversion. One of the most common methods of direct A/D conversion involves the use of a converter than operates on a sequence of *half-split, trial-and-error steps.* This produces code bits in serial form.

The heart of a converter is the *conversion ladder* such as that shown in Fig. 3-34. The ladder provides a means of implementing a 4-bit binary coding system and produces an output that is equivalent to switch positions. The switches can be moved to either a 1 or a 0 position, which corresponds to a four-place binary number. The output voltage describes a percentage of the full-scale reference voltage, depending on the switch positions. For example, if all switches are at 0 position, there is no output voltage. This produces a binary 0000, represented by 0 V.

If switch A is at 1 and the remaining switches are at 0, this produces a binary 1000 (decimal 8). Since the total in a 4-bit system is 16 (0 to 15), 8 represents one-half of full scale. Thus, the output voltage is one-half of the full-scale reference voltage. This is done as follows.

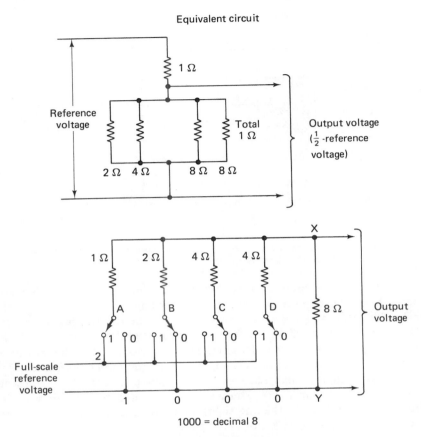

Figure 3-34 Binary conversion ladder used in the 4-bit system.

The 2-Ω, 4-Ω, and 8-Ω switch resistors and the 8-Ω output resistor are connected in parallel. This produces a value of 1-Ω across points X and Y. The reference voltage is applied across the 1-Ω switch resistor (across points Z and X) and the 1-Ω combination of resistors (across points X and Y). In effect, this is the same as two 1-Ω resistors in series. Since the full-scale reference voltage is applied across both resistors in series and the output is measured across only one of the resistors, the output voltage is one-half of the reference voltage.

In a practical converter, the same basic ladder is used to supply a comparison voltage to a comparison circuit, which compares the voltage to be converted against the binary-coded voltage from the ladder. The resultant output of the comparison circuit is a binary code representing the voltage to be converted.

The mechanical switches shown in Fig. 3-34 are replaced by electronic switches, usually containing FFs. When the switch is "on," the corresponding ladder resistor is connected to the reference voltage. The switches are triggered by four pulses (representing each of the 4 binary bits) from the system clock. An enable pulse is used to turn the comparison circuit on and off, so that as each switch is operated, a comparison can be made of the 4 bits.

Typical A/D operating sequence.   Figure 3-35 is a simplified block diagram of an A/D converter. Here, the reference voltage is applied to the ladder

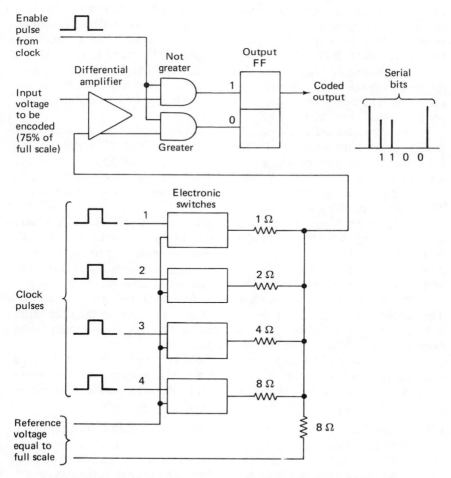

Figure 3-35 Analog-to-digital converter (binary encoder) using the 4-bit system.

through the electronic switches. The ladder output (comparison voltage) is controlled by switch positions which, in turn, are controlled by pulses from the clock, as shown.

The following paragraphs outline the sequence of events necessary to produce a series of 4 binary bits that describe the input voltage as a percentage of full scale (in one-sixteenth increments). Assume that the input voltage is three-fourths of full scale (or 75 percent).

When pulse 1 arrives, switch 1 is turned on and the remaining switches are off. The ladder output is a 50 percent voltage that is applied to the differential amplifier. The balance of this amplifier is set so that its output is sufficient to turn on one AND gate and turn off the other AND gate, if the ladder voltage is greater than the input voltage. Similarly, the differential amplifier will reverse the AND gates if the ladder voltage is not greater than the input voltage. Both AND gates are enabled by the pulse from the clock.

In our example (75 percent of full scale), the ladder output is less than the input voltage when pulse 1 is applied to the ladder. As a result, the *not greater* AND gate turns on, and the output FF is set to the 1 position. Thus, for the first of the 4 bits, the FF output is 1.

When pulse 2 arrives, switch 2 is turned on, and switch 1 remains on. Both switches 3 and 4 remain off. The ladder output is now 75 percent of the full-scale voltage. The ladder voltage equals the input voltage. However, the ladder output is still not greater than the input voltage. Consequently, when the AND gates are enabled, the AND gates remain in the same condition. Thus, the output FF remains at 1.

When pulse 3 arrives, switch 3 is turned on. Switches 1 and 2 remain on, while switch 4 is off. The ladder output is now 87.5 percent of the full-scale voltage and is thus greater than the input voltage. As a result, when the AND gates are enabled, they reverse. The not-greater AND gate turns off, and the greater AND gate turns on. The output FF then sets to 0.

When pulse 4 arrives switch 4 is turned on. All switches are now on. The ladder is now maximum (full scale) and thus is greater than the input voltage. As a result, when the AND gates are enabled, they remain in the same condition. The output FF remains at a 0.

The 4 binary bits from the output are 1, 1, 0, and 0, or 1100. This is a binary 12, which is 75 percent of 16. In a practical converter, when the fourth pulse has passed, all the switches are reset to the off position. This places them in a condition to spell out the next 4-bit binary word.

D/A conversion. A D/A converter performs the opposite function of the A/D converter just described. The D/A converter produces an output voltage that corresponds to the binary code. As shown in Fig. 3-36, a conversion ladder is also used in the D/A converter. The output of the conversion

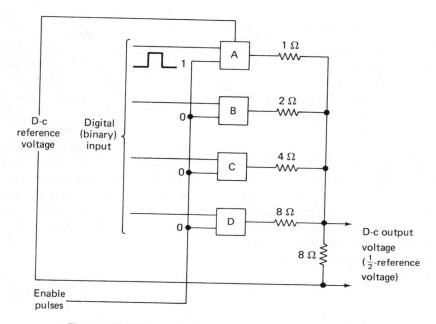

Figure 3-36   Digital-to-analog converter using the 4-bit system.

ladder is a voltage that represents a percentage of the full-scale reference voltage. The output percentage depends on switch positions. In turn, the switches are set to on or off by corresponding binary pulses. If the information is applied to the switch in four-line (parallel) form, each line can be connected to the corresponding switch. If the information is in serial form, the data must be converted to parallel by a register (shift and/or storage).

The switches in the D/A converter are essentially a form of AND gate. Each gate completes the circuit from the reference voltage to the corresponding ladder resistor when both the enable pulse and binary pulse coincide.

Assume that the digital number to be converted is 1000 (decimal 8). When the first pulse is applied, switch A is enabled and the reference voltage is applied to the 1-Ω resistor. When switches B, C, and D receive their enable pulses, there are no binary pulses (or the pulses are in the 0 condition). Thus, switches B, C, and D do not complete the circuits to the 2-, 4-, and 8-Ω ladder resistors. These resistors combine with the 8-Ω output resistor to produce a 1-Ω resistance in series with the 1-Ω ladder resistance. This divides the reference voltage in half to produce 50 percent of full-scale output. Since 8 is one-half of 16, the 50 percent output voltage represents 8.

3-3-2 High-speed A/D converters

Although there are a number of A/D converter schemes, there are only three basic types: parallel, serial, and combination. In parallel, all bits are converted simultaneously by many circuits. In serial, each bit is converted in sequence, one at a time. The combination A/D conversion includes features of both types. Generally, parallel is faster but more complex than serial. The combination types are a compromise between speed and complexity.

Parallel (flash) A/D. Figure 3-37 shows the basic parallel (or flash) A/D conversion circuit, where all bits of the digital representation are determined simultaneously by a bank of voltage comparators. For N bits of binary information, the system requires $2^N - 1$ comparators, and each comparator determines one LSB level. This requires a great number of circuits. Another disadvantage of parallel A/D is that the output of the comparator bank is not directly usable information. The output must be converted to binary information using a decoder.

Tracking A/D. Figure 3-38 shows the basic tracking A/D conversion circuit. Tracking A/D derives its name from the fact that the digital output

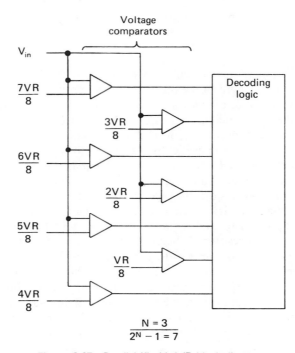

Figure 3-37 Parallel (flash) A/D block diagram.

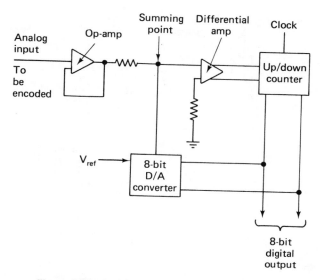

Figure 3-38   Basic tracking A/D conversion circuit.

continuously "tracks" the analog input voltage. Tracking A/D is usually used in communications systems or some similar application where the input is a continuously varying signal. As shown, tracking A/D uses a D/A in a feedback path. The accuracy of the system is no better than the D/A being used (typically 6 to 10 bits).

Successive approximation A/D.   Figure 3-39 shows the basic successive approximation A/D conversion circuit. Note that this circuit is essentially the same as the basic A/D converter of Fig. 3-35. The D/A block of Fig. 3-39 represents the electronic switches and ladder of Fig. 3-35. The successive approximation (S/A) storage register of Fig. 3-39 represents the logic AND gates and the FF of Fig. 3-35. However, 4 bits are shown in Fig. 3-35, whereas 8 bits are shown in Fig. 3-39.

The S/A type of A/D is relatively slow compared to other types of high-speed A/Ds, but the low cost, ease of construction, and system operational features more than make up for the lack of speed (in most applications). S/A is by far the most widely used A/D system. For that reason, we describe a complete S/A type of A/D, using IC components, in Sec. 3-3-3.

As shown in Fig. 3-39, the system enables the 8 bits of the D/A, one at a time, starting with the MSB. As each bit is enabled, the comparator gives an output signifying that the input signal is greater, or not greater, in amplitude than the output of the D/A. If the D/A output is greater than the input signal, the bit is "reset" or turned off. The system does this with the MSB first, then the next most significant bit, then the next, and so on. After

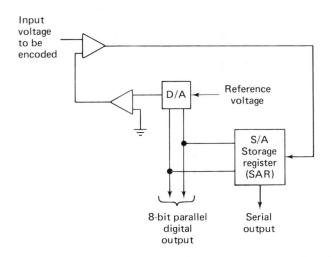

Figure 3-39 Basic successive approximation A/D conversion circuit.

all 8 bits of the D/A have been tried, the conversion cycle is complete and another conversion cycle is started.

The serial output of the system is taken from the output of the comparator. While the system is in the conversion cycle, the comparator output is either 0 or 1, corresponding to the digital state of the respective bit. In this way, the S/A type of A/D gives a serial output during conversion and a parallel output between conversion cycles.

3-3-3 Successive approximation A/D converter

Figure 3-40 shows a schematic diagram of a successive approximation A/D such as that described in Sec. 3-3-2. In the Fig. 3-40 circuit, the basic components include an IC D/A and SAR (successive approximation register). The system requires a total of four ICs. As shown, the system operates on $+5$- and -15-V power supplies, requires approximately 200 mW of power, and operates at 2-μs/bit conversion rates.

In operation, the input voltage V_{in} drives an op-amp A1 connected as a noninverting, unity-gain buffer. For further information on op-amps of all types, refer to the author's *Manual for Operational Amplifier Users* (Reston, Va.: Reston Publishing Company, 1976). The output of the D/A is a current proportional to the reference current I_{ref}, and the digital word on the address lines of the D/A (inputs A1-A8). The voltage on the output of the D/A (V_0) is a function of V_{in} and the output current of the D/A.

The comparator A2 compares V_0 to V_{offset}. If V_0 is greater than V_{offset}, the comparator output is a logic 1. Full-scale voltage (11111111) of the

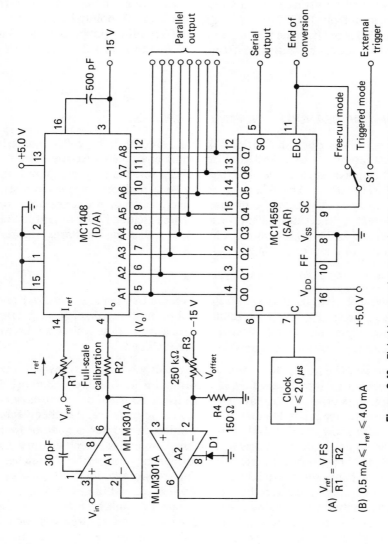

Figure 3-40 Eight-bit successive approximation A/D circuit using ICs.

(A) $\dfrac{V_{ref}}{R1} = \dfrac{VFS}{R2}$

(B) $0.5\ mA \leqslant I_{ref} \leqslant 4.0\ mA$

system is 2.56 V. This gives each LSB a value of 10 mV. Any value of full scale can be chosen as long as the input buffer amplifier does not saturate. The system is calibrated by putting a voltage of full scale, minus ½ LSB, into the input, and adjusting the full-scale calibrate potentiometer R1 to make the transition from 11111110 to 11111111 occur at this point. Then an input of + ½ LSB is put into the system, and the offset adjust potentiometer R3 is adjusted to make the transition from 00000000 to 00000001 occur at this point.

3-4 DIGITAL READOUTS (NUMERIC DISPLAYS)

This section describes the various digital readouts, also called numeric displays, commonly used in digital electronic equipment. At one time, the 10-digit displays, such as the old Nixie® tubes, were the most common form of numeric readout. These have generally been replaced by the seven segment display and will not be discussed here. Instead, we concentrate on readouts using the seven segment format. These include LCDs (liquid crystal displays) and LEDs (light-emitting diodes), as well as gas discharge, fluorescent, and incandescent displays.

3-4-1 Basic digital readout systems

There are two basic types of digital readout systems: *direct drive* and *multiplex*. Generally, it is more economical to multiplex displays of greater than four digits. Thus, multiplexing is emphasized in the following discussions. Several examples of time sharing are included.

Direct-drive displays. The simplest type of display system, shown in Fig. 3-41, consists of four lines of BCD information feeding a decoder/driver whose outputs drive the display. This direct-drive system does not have information storage capability and thus reads out in *real time*. Another display system, also shown in Fig. 3-41, contains a decade counter, a quad or four-line latch (FFs), a decoder/driver, and the display, one such channel for every digit. This alternate system has storage capability (the FF latches), which allows the counter to recount during the storage time.

Both systems have decoder/drivers which convert the BCD count into voltages suitable for operating the seven-segment displays. Figure 3-41 also shows the relationship between the decoder and numerical display, as well as a truth table. As shown by the truth table, the segments are illuminated in accordance with the decimal number applied at the BCD input. For example, for a decimal 3, the BCD signal is 0011, and segments *a, b, c, d,* and *g* are illuminated. Segments *e* and *f* are not illuminated, and the display forms a numeral 3.

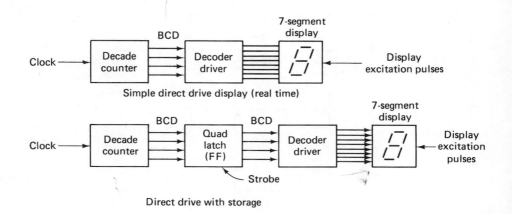

Simple direct drive display (real time)

Direct drive with storage

Relationship between decoder and numerical display

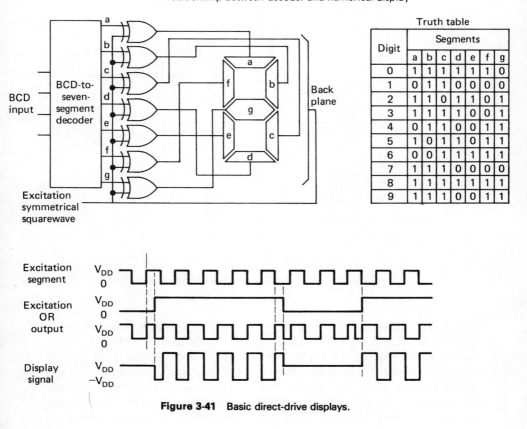

Truth table

Digit	Segments						
	a	b	c	d	e	f	g
0	1	1	1	1	1	1	0
1	0	1	1	0	0	0	0
2	1	1	0	1	1	0	1
3	1	1	1	1	0	0	1
4	0	1	1	0	0	1	1
5	1	0	1	1	0	1	1
6	0	0	1	1	1	1	1
7	1	1	1	0	0	0	0
8	1	1	1	1	1	1	1
9	1	1	1	0	0	1	1

Figure 3-41   Basic direct-drive displays.

159

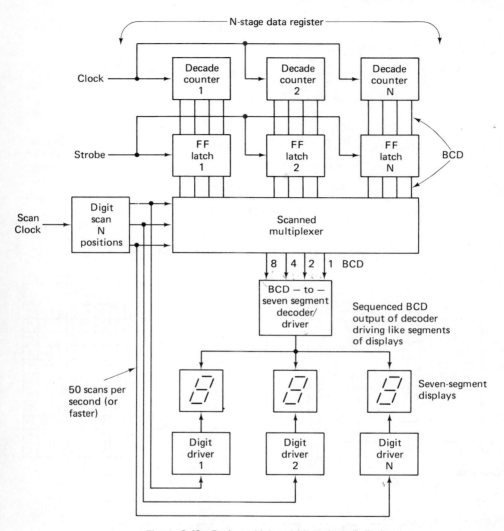

Figure 3-42   Basic multiplexed (time-shared) display.

Multiplex displays.   The most commonly used system for multidigit displays is the multiplexed (or time-shared or strobed) system shown in Fig. 3-42. By time sharing the one decoder/driver, the parts count, interconnections, and power can be saved. The N-stage data register (one stage for each digit) feeds a scanned multiplexer. In turn, the sequenced BCD output of the multiplexer drives like segments of the display. The digit-select elements are sequentially driven by the scan circuit, which also synchronously drives the multiplexer. Thus, each display is scanned or strobed in synchronism, with the BCD data presented to the decoder at a sufficiently high rate, usually greater than 50 scans per second, to appear as a continuously energized multidigit display.

3-4-2 Liquid crystal displays (LCD)

Liquid crystals are fluids that flow like a liquid but have some of the optical characteristics of solid crystals. LCDs consist of certain organic compounds whose characteristics change state when placed in an electric field. Thus, images can be created according to predetermined patterns. Since no light is emitted or generated, very little power is required to operate LCDs. Thus, they are well within the drive capabilities of MOS ICs and are well suited for battery operation. LCDs show excellent readability in direct sunlight. However, in low-ambient-light conditions, some form of light source (either within or external to the display) should be used.

There are two types of LCDs: *dynamic scattering* and *field effect.* In *dynamic scattering LCDs,* the electric field rearranges the normally aligned molecules to scatter the available light. This causes the display to change from a transparent state to an opaque state. The *field-effect LCDs* consist of two pieces of plate glass separated by a glass spacer/seal with the liquid-crystal material injected between the two plates. A metalization pattern is etched onto the glass (typically in the seven-segment numerical display format). The glass and liquid crystal material under the selected segment is activated when a field is placed on that particular segment. Polarizers are attached to the front and back of the display. The light striking the lower polarizer is either reflected or absorbed, depending on the relative direction of the polarizers.

If a *reflective display* is desired, reflective material is adhered to the back polarizer. Reflective displays are generally used in applications where ambient light is available. The light enters the front of the display, goes through the front polarizer and glass, then through the crystal material and back polarizer, reflects off the reflector, and finally goes back out through the front of the display. Reflection operation thus uses external light energy and requires virtually no power.

A *backlighted transmissive display* is often used in environments where only a small amount of light is available. The transmissive display does not use reflective material. Instead, backlighting is done with diffused incandescent or fluorescent light sources.

LCDs require an a-c drive signal, with no d-c component. For field-effect displays, the excitation signal can be as low as about 2 V, typically at frequencies of 60 Hz to 10 kHz. The excitation signal for dynamic scattering displays in the 7- to 30-peak range, at 200 to 400 Hz.

With multiplexed circuits, LCDs ultimately reach their maximum contrast when they are continually pulsed with a low-duty cycle waveform, even when the pulse width is much less than the display turn-on time. This integrating effect is similar to that of a capacitor being charged by a series of pulses. The total time required to reach full contrast is in the range 100 to 400 ms.

The excitation voltage is less for field-effect LCDs than for dynamic scattering LCDs. This results in a corresponding lower *threshold voltage* (the voltage at which the displays switch from off to on). When the duty cycle of the excitation is varied, the effective threshold varies inversely. Thus, as the number of multiplexed LCD ditits increase (duty cycle decreases), the threshold, and therefore the required power supply, must also increase. For dynamic scatterning LCDs, the required power supply is beyond the range of the CMOS ICs.

Example of multiplexed LCD display.   Figure 3-43 shows a 3½-digit LCD multiplexer which incorporates a Motorola MLC401 readout. Figure 3-43 also shows the waveforms for the circuit. The three-digit BCD counters (and respective latches), and the multiplexer and scan circuitry, are all contained in one CMOS Package, designated as the MC14553. The time-division multiplexed BCD outputs are fed to the inputs of the MC14543 BCD-to-seven-segment latch/decoder/driver for liquid crystals. For other types of display (LED, incandescent, and so on), the digits can be strobed by simply scanning the digit drivers. Thus, all common digit segments can be tied to their respective decoder outputs. However, for LCDs, this approach cannot be used, since the display will always see a signal across it, even when the backplane (digit select) is not strobed.

When gates are placed in series with the segment lines, the display can readily be strobed. For this example, MC14001 two-input NOR gates are used, with the control inputs being the negative-going digit select pulses from the MC14553 scan circuit. During the nonselect time when the scan output is high, the outputs of the NOR gates are low regardless of the state of the excitation signal. Zero voltage is then present across the MCL401 display, as shown by the waveforms of Fig. 3-43. During the select low level, the gate outputs are controlled by a decoder output and an excitation signal. The selected digit then operates as a direct-driven system.

The excitation frequency of 100 Hz for the field-effect LCD is derived from the two-CMOS gate (G1 and G2) FF oscillator. This toggle FF square-wave circuit ensures that the excitation signal has an exact 50 percent duty cycle with no d-c component. The scan frequency, using capacitor C_S and the MC14553 internal oscillator, is set for approximately 500 Hz, resulting in a display refresh rate of approximately 170 Hz (500/3), a frequency well beyond the detectable flicker rate of 30 to 60 Hz. The effective threshold

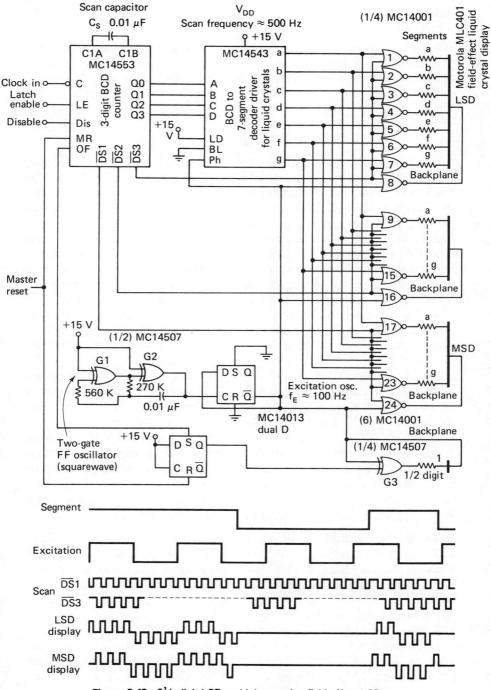

Figure 3-43 3$\frac{1}{2}$-digit LCD multiplexer using field-effect LCDs.

voltage for a three-digit scan (33 percent duty cycle) is approximately 6.5 V. The use of a 15-V power supply ensures that the device is driven well above threshold, with the resulting LCD voltage being approximately 30 V peak to peak.

3-4-3 Light-emitting diode (LED)

LED displays have taken over an appreciable part of the display business in the last several years, primarily due to their high reliability, long life, fast response time, operation from low voltage dc, and ready adaptability to miniature displays. LEDs are semiconductor PN junction diodes which produce light due to recombination of holes and electrons when forward-biased. The semiconductor material is either gallium arsenide phosphide (GaAsP) or gallium phosphide (GaP), with the former being more prevalent in red display applications. Similar to any junction diode, the voltage drop across the forward-biased junction is relatively constant (approximately 1.6 V for GaAsP) and is generally driven with a constant-current source.

There are two possible connections when LEDs are used as seven-segment displays: *common cathode* and *common anode. Common-cathode* displays require the drive circuit to supply current (source) to the segments. *Common-anode* displays require segment drive circuit sink capability (the drive circuit must dissipate the current). Because of their low voltage and relatively small current requirements, LED displays can be readily interfaced with most families of ICs. When such ICs lack the drive capability, transistors can be easily interfaced between the ICs and the display.

Typical GaAsP currents are in the range 5 to 30 mA per segment, generally varying with the size of the display. The light output and luminous efficiency (light output/unit current) increase with increasing forward current until saturation occurs (100 to 150 mA range). These properties make it advantageous to strobe LEDs so that, for the same average current with a lower duty cycle, the peak current and light output are greater.

Example of LED display.   Figure 3-44 shows a typical five-digit real-time LED display system. This system uses the MC14543 BCD-to-seven-segment decoder, and the MC14534 five-digit counter, as the basic control elements. The system also uses IC drivers (MC75591 and MC75492) to drive the Hewlett-Packard HP5082-7740 LED displays. All like anode segments of the common-cathode displays are driven by the emitter outputs of the MC75491. The cathode elements are driven by the MC75492.

The digits are selected, in turn, when an appropriate drive signal is applied to the corresponding common cathodes by the five-digit counter (through the MC75492 drivers). Each digit is selected in order when the appropriate number of scanner pulses are applied to the counter. The counter

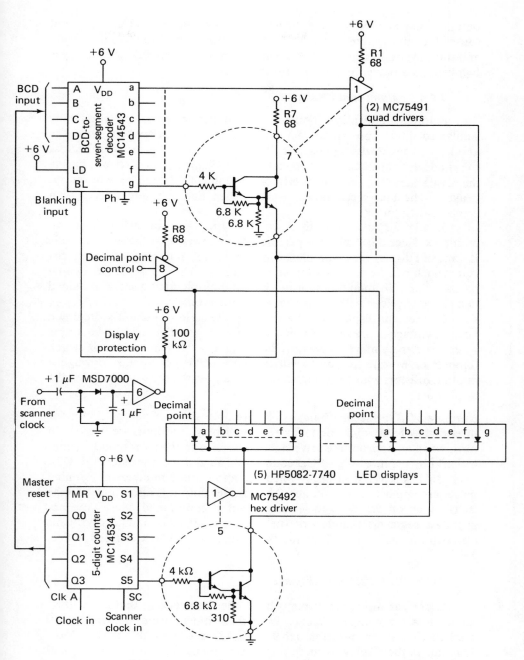

Figure 3-44 Five-digit real-time LED display system.

output is also applied to the BCD input of the seven-segment decoder, and causes the appropriate segments (LEDs) of the selected digit to illuminate and form the desired numeral. Note that if the scan signal fails the display is blanked, thus preventing damage to the displays.

3-4-4   Fluorescent displays

Fluorescent displays, both diode and triode types, are electrically similar to diode and triode vacuum tubes, the exception being that the anodes are coated with a phosphor. When a positive voltage is placed across the anode and the directly heated cathode (or filament), the electrons hitting the anode cause the phosphor to fluoresce and emit light. The light output peaks in the blue-green range, which, when appropriately filtered, can display other colors.

Triode displays with their control grids are somewhat easier to multiplex, since the strobing is performed at a lower power level. The diode fluorescent displays are multiplexed in the relatively high power filament circuit, which can be powered by either ac or dc. The display digits can be packaged in individual tubes, or in a planar multiple-digit display contained in a single envelope with all like segments interconnected.

Fluorescent displays can be energized by series- or shunt-switching a positive voltage to the anode (with respect to the cathode). *Series switching* is when the display anode is energized through a series semiconductor switch (bipolar or MOS). In *shunt switching,* the voltage is applied through a limiting resistor, with the energized switch (across the display) clamping the display off.

Example of fluorescent displays.   Figure 3-45 shows a partial circuit for three digits of a six-digit triode fluorescent display. The digits are selected, in turn, when an appropriate drive signal is applied to the corresponding control grid by the three-digit BCD counter (through drive transistors Q1–Q3). Each digit is selected in order when the appropriate number of scanning pulses are applied to the counter. The counter output is also applied to the BCD input of the BCD-to-seven-segment decoder, and causes the appropriate segments (anodes) of the selected digit to be illuminated. This operation is similar to that of the five-digit LED display described in Sec. 3-4-3.

3-4-5   Gas discharge displays

Older gas discharge displays have one drawback in that all of their numerals are not in one plane. With these older displays, there are 10 preshaped cathodes (numerals 0 to 9) physically stacked within the envelope. This causes the display to jump in and out as the digits are changed, and limits the useful angle of view.

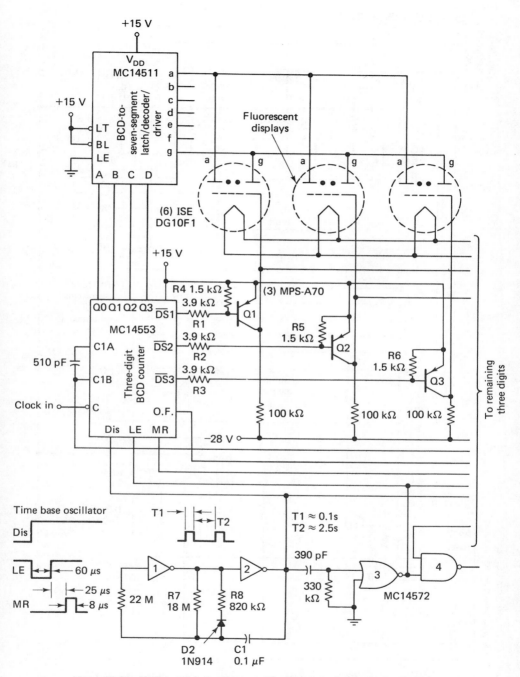

Figure 3-45 Partial circuit for three digits of a six-digit triode fluorescent display.

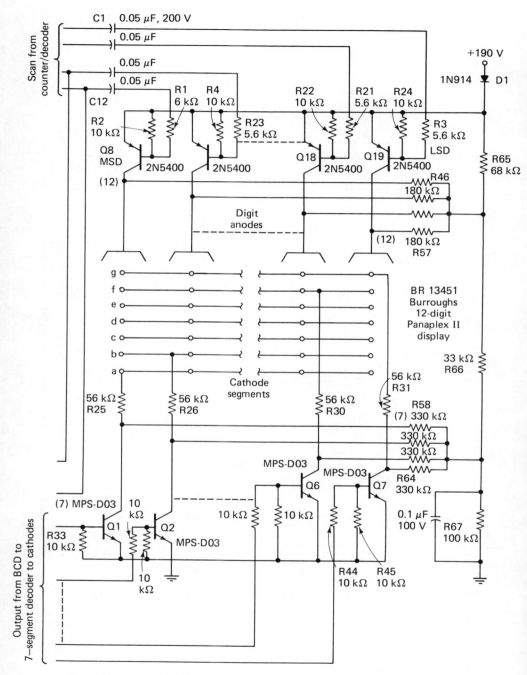

Figure 3-46 Partial circuit for four digits of a 12-digit planar gas discharge display.

The more recently introduced *planar gas discharge displays* have their numerals in one plane, facilitating wide viewing angles. Typically, up to 16 digits can be contained in one neon-filled envelope. Each digit has one anode and seven (or more) cathode segments (to form the seven-segment numeral). The like segments can be tied together as in most multidigit displays or can be brought out individually (if the number of digits is small).

When a voltage greater than the ionization potential (typically about 170 V) is applied between the selected anode and cathode, the gas ionizes and an orange glow appears around the cathode (forming the numeral). For multiplexed displays, a blanking period is required between the cathode select and the anode scan pulses. This ensures that the previous digit is completely deionized before the following digit is strobed and thus prevents erroneous readouts.

Since gas discharge displays require high ionization potentials, interfacing from low-voltage digital ICs to this higher voltages requires level translation or level shifting (as discussed in Sec. 3-5). This can be done by translating upward to the anode drivers, with the cathode drivers referenced to the digital voltages. As an alternative, you can translate downward to the cathode drivers, with the anode circuits at the digital voltage level. Level shifting can be done by directly coupled high-voltage transistors, or by capacitor coupling.

Example of gas discharge displays. Figure 3-46 shows a partial circuit for four digits of a 12-digit planar gas discharge display. The digits are selected, in turn, when an appropriate voltage is applied to the corresponding digit anode from a scanner (counter/decoder) circuit (through level-shifting transistors Q8–Q19). Each digit is selected in order. A BCD-to-seven-segment decoder output is applied (through level-shifting transistors Q1–Q7) to the appropriate segments (cathodes) of the selected digit to be illuminated.

3-4-6 Incandescent displays

The smaller direct-view incandescent displays have seven helical coil segments (to form the seven-segment numeral) fashioned from tungsten alloy. The power requirements for incandescent displays (typically 1.5 to 5 V, at about 8 to 24 mA) make incandescent compatible with LED drivers or decoder/drivers. However, when multiplexing incandescent displays, greater filament peak power is required to maintain a brightness equivalent to LEDs. Also, blocking diodes are required, one for each segment, to provide erroneous display indications through sneak electrical paths.

Example of incandescent displays. Figure 3-47 shows a circuit for a four-digit incandescent display. The digits are selected, in turn, when appropriate scan pulses are applied (through the MC14011 NAND gates

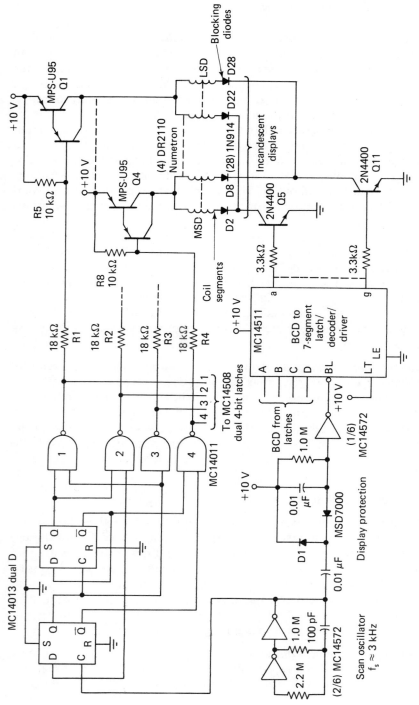

Figure 3-47 Typical four-digit incandescent display.

and transistors Q1–Q4) to all coil segments of the selected digit. The BCD-to-seven-segment decoder output is applied (through transistors Q5–Q11 and diodes D2–D28) to the appropriate segments of the selected digit to be illuminated. Note that if the scan signal fails, the display is blanked, thus preventing damage to the displays. Diode D1 converts scan oscillator pulses into a d-c voltage. This voltage is applied through the inverter to the blanking (BL) input of the BCD-to-seven-segment decoder. If the scan oscillator stops, the d-c voltage developed by D1 is removed and the display is blanked.

3-5 INTERFACING DIGITAL EQUIPMENT

This section describes the various interfacing techniques found in digital electronic equipment. Here, we concentrate on explanations of the basic principles and techniques for interfacing between digital circuits, particularly between the digital IC families. In Chapter 4, we discuss interfacing complete digital equipments such as microcomputers.

No matter what drive and load characteristics are involved for a digital IC, it may be necessary to provide the necessary drive current, to change the logic levels, and so on with an interfacing circuit or device. It is not practical to have a universal interfacing circuit for all families of all digital IC manufacturers. It is not even possible to have a universal circuit for interface between a given family of one manufacturer and all other families of the same manufacturer. Nor it is possible to have a single interface circuit that will accommodate the same digital family of different manufacturers. For one thing, the digital voltage levels (for 0 and 1), the supply voltages, and the temperature ranges vary with manufacturers, and with digital families. For this reason, most digital IC manufacturers publish interfacing data for their particular lines.

We will make no attempt to duplicate all of these data here. Instead, we concentrate on the interface requirements for the most popular lines, and the most used digital families. A careful study of this information should provide you with sufficient background to understand the basic problems involved with interfacing digital systems, and to interpret interfacing data that appears on digital IC datasheets.

3-5-1 Basic interfacing circuit

Figure 3-48 shows the basic circuit for interfacing between digital ICs. The equations shown in Fig. 3-48 are used to find the approximate or trial value of pull-up resistor R1. The following is an example of how to use the equations.

Assume that the common-collector circuit is used, the V_{CC} is 5 V, the V_{OH} (high or logic-1 state voltage) is 3 V, there are seven gate inputs each

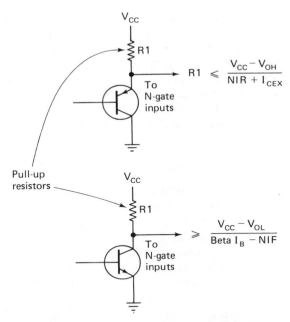

Figure 3-48   Basic circuit for interfacing between digital ICs.

with a current of 1 mA, and that I_{CEX} is 3 mA. Note that I_{CEX} is the input leakage current. Using the equation of Fig. 3-48, the value of R1 is

$$R1 = \frac{5 - 3 \text{ V}}{(7 \times 1 \text{ mA}) + 3 \text{ mA}} = \frac{2 \text{ V}}{10 \text{ mA}} = 200 \text{ } \Omega$$

The next lowest standard value is 180 Ω.

3-5-2   COS/MOS — TTL/DTL interface

The following notes apply to RCA COS/MOS, which is a form of CMOS. When interfacing one digital IC family with another, attention must be given to logic swing, output drive, d-c input current, noise immunity, and speed of each family. Figure 3-49 shows a comparison for COS/MOS, medium-power TTL, and medium-power DTL.

When a bipolar device is used to drive a COS/MOS device, the output drive capability of the driving device, as well as the switching levels and input currents of the driven device, are important considerations. There are three bipolar output configurations to consider: *resistor pull-up, open collector,* and *active pull-up.*

Devices with *resistor pull-ups* can use the basic interface circuit of Fig. 3-48 when used to deive COS/MOS. Devices with *open collectors* require an

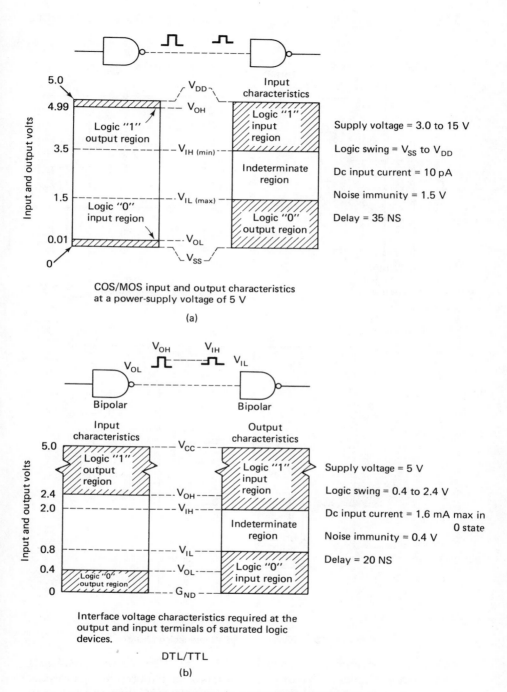

Figure 3-49 Comparison of COS/MOS to TTL/DTL.

COS/MOS input and output characteristics at a power-supply voltage of 5 V

(a)

Supply voltage = 3.0 to 15 V

Logic swing = V_{SS} to V_{DD}

Dc input current = 10 pA

Noise immunity = 1.5 V

Delay = 35 NS

Interface voltage characteristics required at the output and input terminals of saturated logic devices.

DTL/TTL

(b)

Supply voltage = 5 V

Logic swing = 0.4 to 2.4 V

Dc input current = 1.6 mA max in 0 state

Noise immunity = 0.4 V

Delay = 20 NS

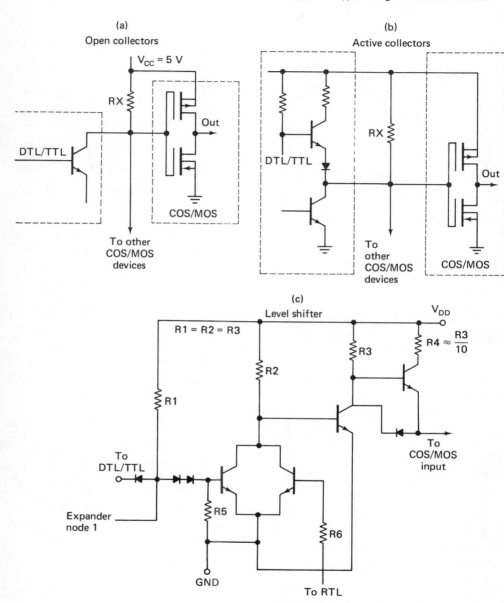

Figure 3-50 Typical COS/MOS interfacing circuits.

external pull-up resistor, as shown in Fig. 3-50(a). When an *active pull-up* is used, a circuit similar to that of Fig. 3-50(b) is often required. In many cases, it is necessary to use a *level shifter* or *level translator* IC such as is shown in Fig. 3-50(c).

From HTL to RTL, DTL, or TTL ($\frac{1}{3}$ of circuit shown)

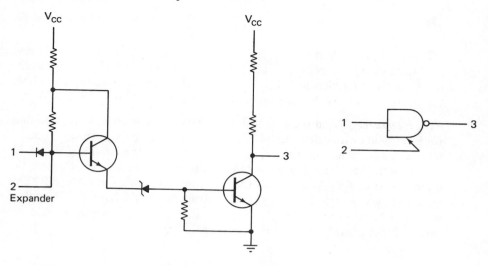

From RTL, DTL, or TTL to HTL ($\frac{1}{3}$ of circuit shown)

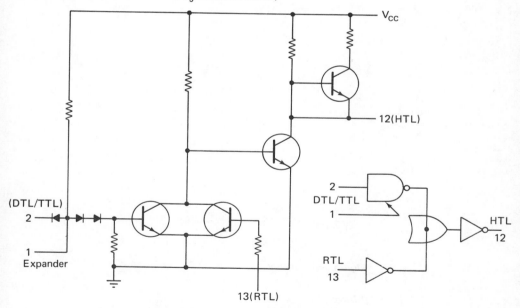

Figure 3-51 Typical HTL interface circuits.

3-5-3 ECL and HTL level translators

In general, level translators (or level shifters) are used when ECL and HTL must be interfaced with any other digital family. This is because of the special nature of ECL and HTL (ECL is nonsaturated logic, and HTL operates with large logic swings).

175

HTL interface. Most HTL logic lines include at least two translators: one for interfacing from HTL to RTL, DTL and TTL; and another for interfacing from these families to HTL. The schematic diagrams and logic symbols for IC versions of such translators are shown in Fig. 3-51.

ECL interface. As in the case of HTL, ECL level translators are available for interfacing with other digital ICs. However, if the only problem is one of interfacing from an input to ECL, it is possible that the circuit of Fig. 3-52 will solve the problem. The input (either discrete component or IC) must be on the order of 1 V. As shown by the equations, the values of R1 and R2 are selected to give a logic 0 (-1.5 V) at the output (to the ECL input). The value of C can be determined by the fact that the RC time constant should be several times greater than the duration of the input pulse.

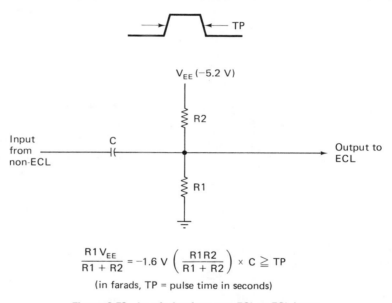

$$\frac{R1\,V_{EE}}{R1 + R2} = -1.6 \text{ V} \left(\frac{R1R2}{R1 + R2} \right) \times C \geq TP$$

(in farads, TP = pulse time in seconds)

Figure 3-52 Interfacing from non-ECL to ECL input.

3-5-4 RTL interface

Because RTL is essentially a discrete component digital family, it is not directly compatible with other (more popular) digital IC families (although it is quite compatible with discrete component devices). The ideal solution is to use level translators. However, because RTL is no longer popular, the available RTL translators are rapidly decreasing. The problem of RTL interface with other digital families can sometimes be solved using simple circuits. Figure 3-53 shows some typical examples for such circuits.

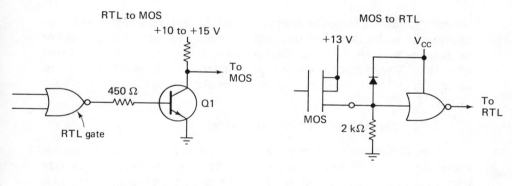

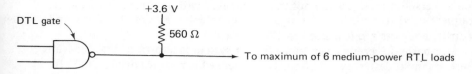

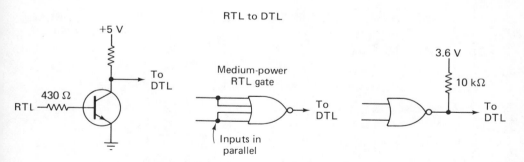

Figure 3-53 Typical RTL interfacing circuits.

3-5-5   Summary of digital IC interface problems

As discussed, each digital IC line or family has its own set of design or usage problems. Most of these problems can be overcome using the datasheets and brochures supplied with the device. One area where the datasheets leave a gap is the interfacing between digital ICs, and to other devices. For that reason, we shall summarize the interfacing problems here.

If MOS is involved, follow the recommendations described in Sec. 3-5-2. If either ECL or HTL are involved, the best bet is to use the various level translators supplied with most ECL and HTL lines, or use the specified translator circuit shown in the datasheets and brochures. In the absence of such information, use the circuits described in Sec. 3-5-3. If RTL must be interfaced, follow the recommendations of Sec. 3-5-4. DTL and TTL are compatible with each other, thus eliminating the need for special translators or interface circuits. However, the load and drive characteristics may be altered somewhat. These changes are generally noted on the datasheets. In the absence of any DTL/TTL interface data, use the information of Secs. 3-5-1 through 3-5-4.

3-5-6 Digital line drivers and receivers

Line drivers and receivers are used where it is necessary to transmit digital data from one system to another. The distances involved may vary from a few feet to several thousand feet. This section summarizes the various characteristics associated with IC line drivers and receivers. (Note that IC line drivers and receivers are not to be confused with modems and acoustic couplers used to transmit digital data over telephone lines. Modems and acoustic couplers are discussed in Chapter 4.)

A basic driver/receiver transmission system is shown in Fig. 3-54(a). Here, an input data stream V_{IN} feeds a driver, which in turn drives a line. Information at the other end of the cable is detected by the receiver that pro-

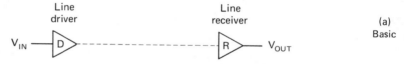

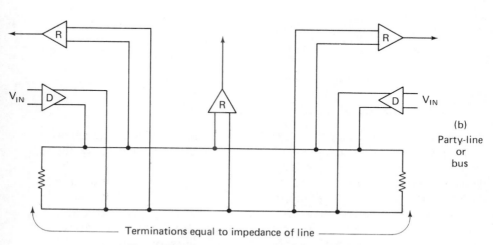

Figure 3-54 Typical digital line driver/receiver systems.

vides an output data stream V_{OUT} usually of the same digital level as V_{IN}. The line involved can be a single line, a coaxial line, a twisted pair line, or a multiline cable (ribbon cable type, multitwisted pair type, and so on). The line can be operated in a *single-ended* mode, but the *differential* mode (requiring a pair of lines) is more popular (because the differential mode is less susceptible to noise problems).

Another common driver/receiver system, shown in Fig. 3-54(b), is commonly called a "party-line" or "bus" system. Here, the line is shared by drivers and receivers. Note that although any driver can be used to drive the line, only one driver is used at one time. The *line driver* translates the input digital levels (TTL, MOS, CMOS, RTL, and so on) to a signal more suitable for driving the line. An important exception to this is found in the ECL family, where ECL gates are often used to drive the line directly. The *line receiver* provides the reverse function of a line driver. With a line receiver, the voltage previously applied to a line is detected and restored to an output level compatible with other digital ICs in the system (TTL, MOS, etc.).

There are ICs designed as line receivers and drivers. Figure 3-55 shows a typical differential line driver/replacing the fixed + 5 V with a strobe pulse. Strobing permits the system to be used in party-line or bus applications. For example, if several drivers (MC75110L) are used to drive the line at different times, each can be enabled when desired (while the other drivers not in use are disabled). Figure 3-55 also shows how line driver/receiver system can be formed using ECL gates.

3-5-7 Optoelectronic interface

Digital ICs are sometimes interfaced with other devices (particularly discrete component circuits) by means of *optoelectronic couplers* or *optocoupler.* An optotelectronic coupler is formed when a LED is packaged with a photodetector as shown in Fig. 3-56. Typically, the LED is of the gallium arsenide, infrared-light-emitting type, whereas the detector can be a single phototransistor or photo-Darlington. When current is applied to the LED, the light emitted by the LED increases current flow through the phototransistor, causing the phototransistor to appear as a short or very low resistance. Both the LED and phototransistor are sealed in a lightproof package, so external light has no effect on operation of the optocoupler.

Optoelectronic couplers are designed for applications requiring electrical isolation and have input/output isolation voltages as great as 2500 V, with isolation resistances being typically in the 10^{11}-Ω range. Owing to the high isolation properties of these couplers, and a typical bandwidth capability from dc to about 300 kHz, optoelectronic couplers can be used where it is required to isolate higher voltage and power loads from low-level digital cir-

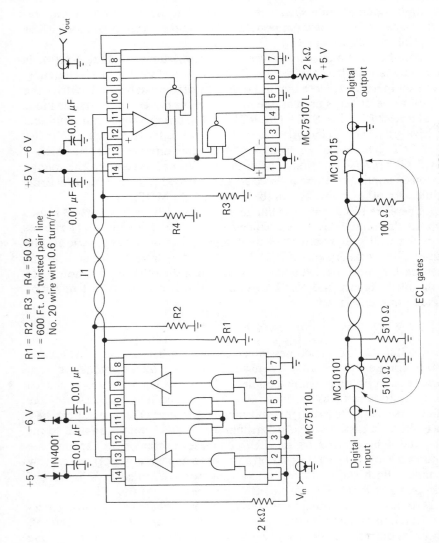

Figure 3-55 Digital line driver/receiver systems using ICs and ECL gates.

R1 = R2 = R3 = R4 = 50 Ω
I1 = 600 Ft. of twisted pair line
No. 20 wire with 0.6 turn/ft

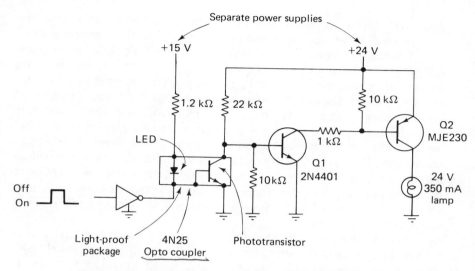

Figure 3-56 Lamp power control circuit using an optocoupler.

cuits. One of the main advantages of the optocoupler is that power supplies can be isolated (separate power supplies for the load and the digital ICs).

In the circuit of Fig. 3-56, a MOS inverter is interfaced with a discrete-component lamp power circuit. The lamp is normally deenergized when the optocoupler is energized. A high (1) level on the input of the MOS inverter energizes the optocoupler, clamping the base–emitter junction of Q1 to off. Transistor Q1 removes drive to Q2, thus deenergizing the load. When a digital 0 is applied to the optocoupler, the clamp to Q1 is removed, and Q2 is driven into conduction through the load, thus energizing the load.

3-6 DIGITAL MEMORY BASICS

This section is devoted to the basics of digital electronic memories. In Chapter 4, we discuss practical examples of the most popular types of digital memories. Here, we concentrate on the basic operating principles of digital memories. In digital electronic literature, the terms *memory* and *storage* are often interchanged. Actually, any device that has a memory is capable of data storage. Before going into how memory and storage circuits operate, let us summarize some common terms used.

Storage can be internal or external. Punch cards, punched paper tape, and magnetic tapes, drums, or disks are some examples of *external storage.* These items are also examples of permanent storage and *nondestructive storage,* since the stored information will remain permanently and is not destroyed during readout. Moreover, the cards and paper tape are *nonerasable,* whereas the magnetic tapes, drums, and disks are *erasable.*

The most common forms of *internal storage* are registers, ferrite cores, and solid-state memories (although magnetic tapes, drums, and disks are considered to be internal in some digital computers). Generally, ferrite cores and solid-state memories are used as *primary storage,* while registers are used for *secondary storage* or for *transfer.* Ferrite cores and some solid-state memories are erasable. Thus, it is necessary to *rewrite data* into them after readout. This is sometimes known as *refreshing the memory.* Although erasable and destructive, ferrite cores and some solid-state memories will retain information when power is removed. Such ferrite cores and solid-state memories have greater *permanence* than registers, and are considered *nonvolatile.* The information in a *volatile memory* is destroyed when the power is removed.

The FFs that make up registers have a memory, since they can be set to a given state and will remain in that state until a specific condition changes them. For example, if an FF receives a 1 input, the FF will set to the 1 state. If another 1 input is applied, the FF will "remember" that it is in the 1 state and will not change. Only a 0 input can change the state. Usually, one FF is used for each bit of data. Thus, a register with 8 FFs has a *storage capacity* of 8.

Ferrite core memories and most solid-state memories have short *access time,* which is the time required to retrieve data from storage. Memories can have different *modes of access.* Some *serial access* or *sequential access,* where each bit is read out one after another. Other memories use *parallel access,* where all bits of a binary word are read out simultaneously. When data can be read out of storage at a particular address, without reference to other addresses, the memory is said to have *random access* or *direct address.* Ferrite core memories and some solid-state memories provide random access. There will be more terms as we go along.

3-6-1 *Punch-card and paper-tape memory systems*

Punch-card data systems were in use many years before digital electronics was developed. These systems were capable of punching data into the cards, sorting or rearranging the cards in any desired order or sequence, and reading the data out (usually on a typewriter or some form of printer). These early punch-card system were essentially electromechanical, and did not use solid-state digital gates. For these reasons, we will not go into punch-card systems here.

Paper-tape reader and punch. A paper-tape reader provides for direct input to a computer or other digital device by reading prepunched data in paper (or thin metallic) tape. A paper-tape punch provides a permanently recorded output from the computer or digital device by punching output information in the paper. Often, the two functions are combined in a single instrument. One common use for a paper-tape instrument is to convert a computer *program* into *machine language*. For example, a program can be entered into a computer using *assembly language* and an *assembler*. The output from this computer is a program in machine language that is punched into paper tape. The punched-paper-tape program is then read into the computer in machine language. (These terms are discussed in Chapter 4.)

Paper-tape machines are also quite useful in computer/data-processing applications, and were in use long before the development of digital electronics. For example, even today, recording on paper tape is often done with machines that punch data received directly from a typewriter or keypunch, or over telephone lines. Paper-tape-reader output can sometimes be used to drive electronic typewriting directly, or over telephone lines.

Paper-tape-reader basics. Figure 3-57 shows the basic elements of a paper-tape reader. As shown, information stored on paper tape is recorded in patterns of round punched holes located in parallel tracks (or channels) along the length of the tape. A character is represented by a combination of punches across the width of the tape. Paper tapes vary in width according to the number of channels they contain. Typically, paper tapes have either five or eight channels, with an eight-channel tape being the most common. Paper-tape systems are ideally suited to the binary (or two-state) system used in digital electronics, since the basic indication of paper tape is either "hole" or "no hole." Two methods of sensing the binary bits in a microcomputer system are shown in Fig. 3-57.

In one system of Fig. 3-57, wire brushes complete a circuit through the tape holes to a metal plate underneath the tape. As the tape is drawn across the plate, the brushes either complete the circuit through a hole (producing an output pulse) or fail to complete the circuit where there is no hole (producing no output pulse). Generally, an output pulse (hole) represents a binary 1, where no output pulse (no hole) represents a 0. One brush is used for each track or channel of holes.

In the other system of Fig. 3-57, a light source is placed on one side of the paper tape, with photocells located on the opposite side. One photocell is used for each track or channel of holes. As the tape moves, the light strikes a photocell wherever there is a hole, producing an output or binary 1. No output (binary 0) is produced where there is no hole, since the light cannot pass to the corresponding photocell.

The holes in the paper tape can be located only at predetermined sites, as shown in Fig. 3-57. Each set of holes across the tape represents one character. A series of characters makes up a word. Location holes guide the tape through the reader and hold the tape in proper position for reading. At

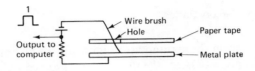

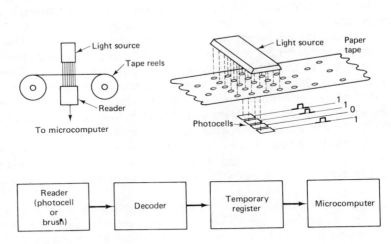

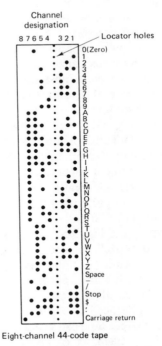

Eight-channel 44-code tape

Figure 3-57 Basic paper-tape reader circuits (From Lenk, *Handbook of Microprocessors, Microcomputers, and Minicomputers,* Prentice-Hall, Inc., 1979).

each character, the tape is stopped momentarily for reading, after which the character is stored in a temporary register. The use of this register between the reader and the microcomputer input compensates for the difference in speeds of the reader and microcomputer. In some systems, the temporary register is part of the microcomputer system. If the paper tape is not punched in machine language (typically an 8-bit binary data byte), a decoder is required between the tape reader and microcomputer input. The decoder converts the five- or eight-channel output from the reader into a format (machine language) suitable for the microcomputer.

Paper-tape punch. When paper tape is used at the microcomputer output, the holes are produced by solenoid-operated metal punches on the paper-tape punch. One punch is used for each track or channel, as shown in Fig. 3-58. The output from the microcomputer is applied through a decoder to the punch solenoids. (If the paper tape is to be punched in machine language, the decoder can be omitted.) The decoder converts the microcomputer-processed information from machine language into a five- or eight-channel code, as applicable.

The paper tape is driven past the punches by the same gear mechanism used in reading the tape. There is one locator hole on the tape for each *frame*

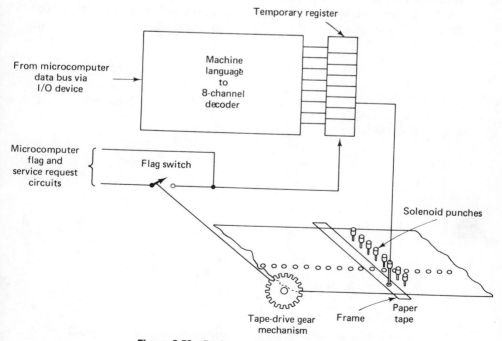

Figure 3-58 Basic paper-tape punch circuits.

(set of holes across the width). As the frame is pulled into position below the punches, a *service request signal* (sometimes known as a *flag* or *flag signal*) is sent to the microcomputer. If a data byte is available at the microcomputer output, a service request permits the data byte to be passed to the punch solenoids. If there is no data byte available, the tape drive is stopped until new information is available. Thus, the microcomputer output is synchronized with the timing of the tape punch. Of course, under normal circumstances, the microcomputer will produce results faster than can be punched by the tape unit. For this reason, a temporary register is often used between the microcomputer and tape unit.

3-6-2 Magnetic recording basics

Magnetic tape, drums, disks, and ferrite-core memories all operate on the principle of *electromagnetism.* Therefore, we shall review electromagnetism as it applies to digital memory of data storage.

As shown in Fig. 3-59, current through a coil creates a magnetic field. If a metal capable of being magnetized is placed in this field, the metal will become magnetized in one direction. If current flow is reversed, the metal will be magnetized in the opposite direction. As shown in Fig. 3-60, if metal is magnetized in one direction, and an opposite magnetizing current is applied to change the direction of magnetism, the changing magnetic field produces (or induces) a current in another conductor (in or near the field).

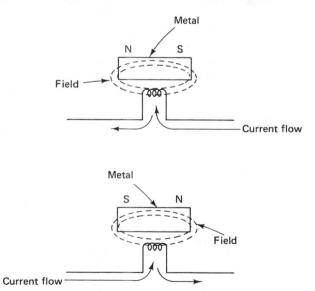

Figure 3-59 Basic principles of magnetic storage, showing metal magnetized in opposite directions by opposite currents.

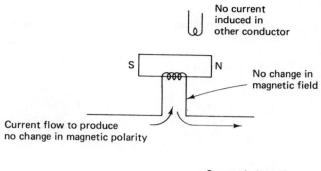

No current
induced in
other conductor

S N

No change in
magnetic field

Current flow to produce
no change in magnetic polarity

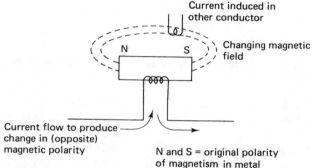

Current induced in
other conductor

Changing magnetic
field

N S

Current flow to produce
change in (opposite)
magnetic polarity

N and S = original polarity
of magnetism in metal

Figure 3-60 Basic principles of magnetic storage or memory, showing the effects of currents on previously magnetized metal storage elements.

However, if the magnetizing current is applied so as not to change the direction of magnetism, there is no change in the magnetic field and no induced voltage. Thus, the magnetized metal has a "memory" and is ideally suited to the two-state binary (1 or 0) system used in digital electronics.

There are two ways in which this magnetic principle can be used. In the case of magnetic tapes, drums, and disks, the *presence* or *absence* of magnetism is used to indicate the binary state. Generally, the presence of magnetism indicates a binary 1, whereas a binary 0 is indicated by the absence of magnetism.

In the case of ferrite cores, the *direction* of magnetism is used to indicate the state. As is discussed in Sec. 3-6-3, ferrite cores are circular and can be magnetized in the clockwise or counterclockwise direction, depending on the direction of current flow. If new currents are applied, the cores will "remember" their existing state and will change states or remain in the existing state, as applicable.

Hysteresis. Whenever metal is magnetized in one direction and an opposite magnetizing force is applied, there is some lag before the magnetic field changes. This lag is known as hysteresis, which can be shown by means of a

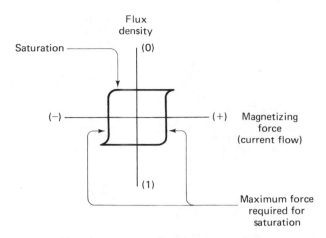

Figure 3-61 Hysteresis graph or loop showing relationship of magnetizing force to flux density in magnetic storage elements.

graph or loop, as illustrated in Fig. 3-61. Such a graph shows the relationship between magnetic force (current in the case of an electromagnet) and magnetic flux density. *Magnetic force* is shown on the horizontal line, with any point to the right of the vertical center line indicating current flow in one direction (arbitrarily labeled +). Points to the left of the vertical center line indicate current flow in the opposite direction (−). *Flux density* is shown on the vertical line. Any point above or below the horizontal line can be considered (arbitrarily) as north or south, clockwise or counterclockwise, and so on. In our example, we shall consider the binary 0 state at being above (with binary 1 below) the horizontal line.

The vertical sides of the hysteresis loop represent the maximum magnetizing force required to produce saturation of the magnetic flux (represented by the horizontal sides of the loop). Note that the loop is rectangular, almost square. This is the ideal condition for magnetic devices used in digital memories. A perfectly square hysteresis loop (rarely found in any magnetic device) means that the device is perfectly linear. That is, a given current in either direction produces the same flux density in the corresponding direction. Next, and particularly important for ferrite core memories, the flux density remains at about the same level when the magnetizing force is returned to 0, and even when an opposite magnetizing force is applied. In fact, it takes about twice the magnetizing force (or current) to change status (1 to 0, or 0 to 1) as is required to go from a demagnetized condition to either state. A square loop device has considerable hysteresis or lag and is thus a good memory. A nonsquare loop magnetic device begins losing flux density when the magnetizing force is removed, and loses density quickly in the presence of an opposite force, making nonsquare devices poor memories.

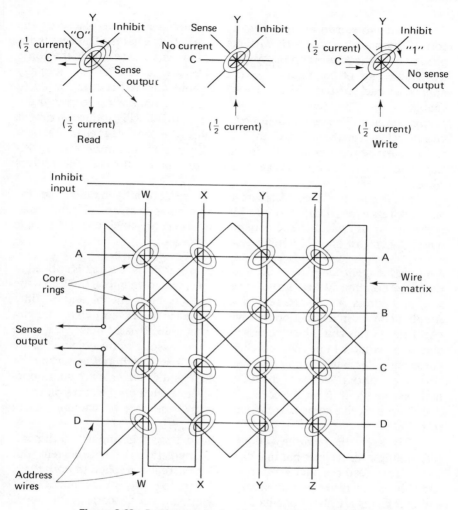

Figure 3-62 Read and write functions in ferrite core memories.

3-6-3 Ferrite-core memories

Ferrite-core memories used in digital electronics consist of metal rings arranged on a wire matrix, as shown in Fig. 3-62. The core rings are made of metal alloys that possess rectangular or square hysteresis-loop qualities. Note that each core has four wires passing through it. The wires labeled ABCD and WXYZ are the *address wires*. Currents passing through the address wires set the state (1 or 0). The core state is dependent on the direction of current flow.

Although shown at right angles, the address wires are placed on the core so that the effects of their currents are additive. That is, if current is applied to both the A wire and the W wire simultaneously, the core at the upper left-hand corner of the matrix receives double current. The currents normally applied to each of the address wires are about one-half that necessary to change states. Thus, current must be *applied to both wires* to produce a change in states. The *sense wire* is positioned so that it will produce an output current whenever there is a change of states (the core magnetism changes from clockwise to counterclockwise, and vice versa). The *inhibit wire* is positioned so that it will oppose the effect of the current through the address wires.

In the matrix of Fig. 3-62 there are 16 cores, one at each junction of the eight address wires. Thus, a 0 or a 1 bit can be stored at any of 16 locations or addresses. In at typical ferrite-core memory, there are 4096 cores on a single matrix, each row and column having 64 cores.

Read/write operation. To *write* data into any address, current is applied to the corresponding address lines. For example, as shown in Fig. 3-62, to write a 1 at address YC, a current is applied to both the Y and C lines simultaneously, in a direction that will magnetize the YC core to the 1 state. (Arbitrarily, this could be clockwise or counterclockwise. Similarly, a write command could be one that sets the core to 0, as long as the commands are consistent. For our example, we consider a write command as setting the core to a 1 state.) Note that the individual Y and C currents are about one-half that required to set the core. Thus, even though all the cores on the Y and C lines receive some current, only the YC core receives enough current to be set.

To *read* data from any address, current is again applied to the corresponding address lines but in a direction opposite to the write current. For example, to read the data at address YC, a current is applied to both the Y and C lines in a direction that sets the YC core to 0. Again, the individual Y and C currents are about one-half that required to set the core. If the core is previously set to 1, the combined YC currents set the core to 0. This changes the direction of magnetism (say from clockwise to counterclockwise), producing a voltage at the output of the sense lines. In practice, the sense voltage is amplified and is used to set an FF in a register. In our example, the presence of a sense line output is considered as a 1, and sets the FF to 1.

If the core is previously set to 0, the combined YC currents have no effect on the core, since the core is at 0. There is no change of magnetic direction and no voltage output from the sense line. The absence of sense-line output is considered as 0, and leaves the FF at 0. (The register FFs have previously been cleared to 0 by a timing pulse.) If the core happens to be demagnetized and in some intermediate state (either 0 or 1), the read currents set the core to 0. However, the small change in magnetic state produces a small sense output voltage (not sufficient to set the register FF).

Note that the cores are arranged on the address wires at angles and that the sense wire passes through the core groups in opposite directions. This arrangement is to minimize the effect of noise produced by the address wire currents. Although these currents are insufficient to set the cores (unless there is coincidence of the two currents at a particular address), the currents can produce some change in the magnetic field and thus produce voltages (or noise) in the sense line. Using the arrangement of Fig. 3-62, the voltages produced by one set of cores is of opposite polarity to the other set of cores, canceling any noise voltage in the sense line.

Avoiding a destructive readout.   The read operation can be destructive. For example, if the core is in the 1 state, a read will set the core to 0. It is then necessary to write the 1 back into the ferrite memory at that address. This can be done by reading the register into which the memory is applied (output of the sense line) and then writing the output of the register back into the memory (by applying currents to the appropriate address wires). Keep in mind that the memory address may have originally been at a 0 when the read currents were applied. In this case, there is no output from the sense line, and the register remains at 0. For each bit where the register is at 0, the write current must be prevented from setting the core to 1. There are several methods for doing this. The most common way is to apply a current through the inhibit line, canceling the effect of the write currents. The inhibit current is not sufficient to change the states of the cores, but is sufficient to nullify a write command.

The inhibit current must be turned on during a write command only when the particular address is to remain at 0. Figure 3-63 shows a typical inhibit control circuit. The bit output from the memory register is connected to an inverter. The output of the inverter is applied to an AND gate together with an inhibit pulse. The inhibit pulse is present during the write command. The inhibit pulse output from the AND gate is amplified and applied to the inhibit line. If the bit read out of the memory into the register is a 1, the 1 is inverted to 0, keeping the AND gate closed. The inhibit pulse is not applied to the inhibit line, and the 1 is rewritten back into memory. If the original memory bit is 0, the 0 is inverted to a 1, opening the AND gate. The inhibit pulse is then applied to the inhibit line, preventing the write command from setting the memory core to 1.

Ferrite memory drive circuits.   The driving currents required for the address lines of a typical ferrite core memory are on the order of 0.5 to 1 A. This amount of current requires amplifiers using power transistors. The amplifiers must be capable of reversing current flow, depending on the com-

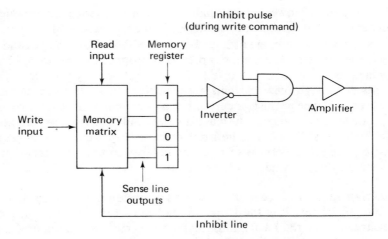

Figure 3-63 Typical ferrite-core memory inhibit control circuit.

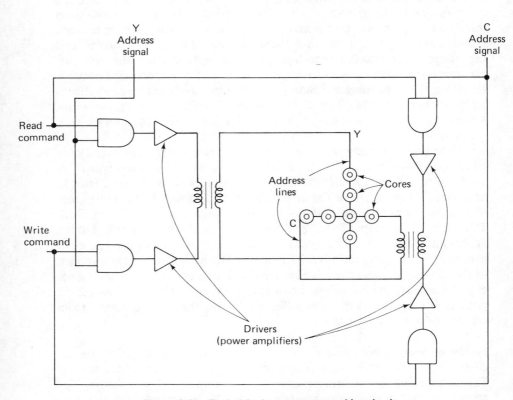

Figure 3-64 Typical ferrite-core memory drive circuit.

mand (read or write). Figure 3-64 shows typical ferrite memory drive circuits. There are two sets of power amplifiers, one for the Y address line and
another for the C line. The amplifier on drivers are placed in a condition to
conduct only when both inputs to either AND gate are true. One set of AND
gate inputs is made true by the Y and C address signals (as selected by the
computer or other program control). The other sets of inputs receive read or
write command pulses. During the write operation, both the Y and C signals
are present, as are write pulses. Current flows through the Y and C address
lines in a direction to set the YC core to 1. During read, the Y and C signals
are still present, as is the read pulse. Current flows through the Y and C lines,
but in the opposite direction, setting the YC core to 0.

Memory words and addresses. There are 16 possible addresses for each bit
on the matrix of Fig. 3-62. One matrix (or one *bit plane*) is required for each
bit used in digital words. That is, if digital words are 4 bits long, 4 bit planes

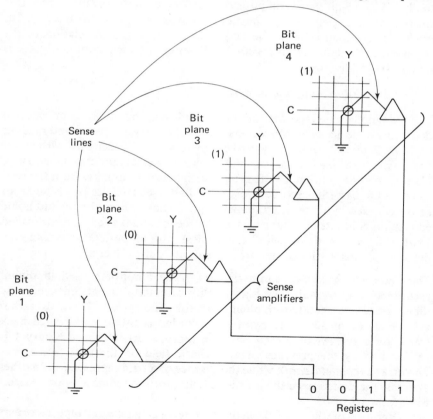

Figure 3-65 Typical ferrite-core memory arrangement for a 4-bit word.

are required, as shown in Fig. 3-65. To read out a complete word into the register, the corresponding address of each bit plane must be read. In the example of Fig. 3-65, the Y and C address lines of all four planes are driven at once. The YC cores of the first and second bit planes are in the 0 state, producing no output from the first and second sense amplifiers. Thus, the first and second positions of the register remain at 0. The YC cores of the third and fourth bit planes are in the 1 state, producing outputs on the corresponding sense amplifiers. Thus, the third and fourth positions of the register are set to 1. This produces a binary 0011 (decimal 3).

In the ferrite core memory of a typical computer, each plane contains 4096 cores or addresses. Sometimes dual planes are used, producing 8192 addresses. In expressing the storage capacity of a computer, both the *number of addresses* (rounded off to the nearest thousand) and the *word length* are given. For example, a typical unit is described as an 8K (eight thousand, actually 8192), 16-bit-word computer. In some computers, the memory is further divided into *pages*. These memory pages are not physical divisions, but represent a system of how the addresses are arranged in relation to the registers. Although there is no standardization, a typical computer page is about 1000 addresses.

3-6-4   Tape recording basics

Magnetic-tape recorders are often used with computers as an external data storage device. These recorders provide high-speed read-in and readout of data. The most common type of magnetic-tape recorder used with larger computers is the *reel-to-reel* system familiar to the data processing industry. *Cassette* recorders are most often used with microcomputers. Both the reel-to-reel and cassette recorders used in digital electronics are similar to recorders used in home entertainment units. That is, both digital and home entertainment units have tape transports, record and playback heads (called *write* and *read* heads in digital electronics), and amplifiers. However, the detailed characteristics of the two systems are quite different.

Tape recorder amplifier characteristics.   The amplifiers used in digital recorders do not require the high fidelity of entertainment units, because digital recorders operate with binary data (bits and bytes of 1s and 0s) rather than voice or music. Data bytes are recorded (written) and played back (read) on a *present* or *absent* basis (in most cases). Often, a binary 1 is represented by the presence of a magnetic field on the tape (sometimes known as a *magnetic spot*), while the absence of a field represents 0. In other systems, the direction of the magnetic field (north or south, + or − , etc.) represents a 1 or 0.

Because of the on–off method of recording, digital tape recorders generally do not have the supersonic bias signal applied to the tape for linearity, as is the case in other recorders, where high fidelity is required. However, digital recording systems often use a fixed d-c bias to place the tape in condition to be magnetized by the data bytes.

Tape characteristics. The tapes used in digital reel-to-reel recorders are generally longer than those of home units; 2400 or 3600 feet is standard. The cassette tapes used with microcomputers are generally the same as for entertainment units. Although early magnetic tapes were made of metal, all popular tapes are now made of plastic, coated on one side with metal oxide. The oxide can easily be magnetized and retains its magnetism indefinitely. The data bits are placed across the width of the tape on parallel tracks running along the entire length. Figure 3-66 shows the recording format for a typical cassette tape used with microcomputers.

Tape coding systems. The pattern of the magnetized spots across the width and along the length of the tape is a coded representation of the data stored on the tape. Several codes are used, as is the case with paper tape. If some code other than machine language has been used to record data on the tape, a decoder is required between the tape reader and the digital equipment input. Similarly, the output from the digital equipment is applied through a decoder to the tape recorder during "record" or "write" operation (unless the data bytes are to be recorded in machine language).

Figure 3-67 shows a tape recording format using the ASCII code. Note that there are eight channels on one track. Seven of the channels are used for the 7-bit ASCII code, and the eighth channel is used for a parity bit. Thus, the eight channels provide for one character (letter, number, or symbol) across the width of the tape. In Fig. 3-67, each dash or short line represents a binary 1. The absence of a dash represents a binary 0.

The data bytes (representing characters) are recorded serially, as the tape passes by the read/write heads. This is shown in Fig. 3-68, which illustrates 1 bit (on one channel) of four successive data bytes. A magnetic field (or spot) is impressed on the tape (representing a binary 1 bit) as current (in the form of pulses) passes through the "write" winding. (In some systems, one head is used for both read and write. In other recording systems, separate read and write heads are used). In the absence of current, no spot is impressed on the tape. During the read operation, as the tape is passed across the head by the tape transport, the presence of a moving magnetic spot produces current in the read head (for a binary 1 bit). No current is produced when there is no spot on the tape passing the read (binary 0 bit).

The system shown in Fig. 3-68 uses the return-to-zero (RZ) format. Tapes are also recorded in NRZL and NRZM (discussed in Sec. 3-3 and illustrated in Fig. 3-32). The RZ format is the least efficient for storing the

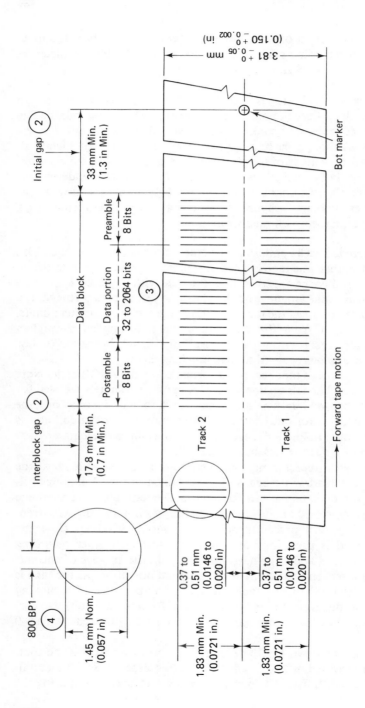

Figure 3-66 Recording format for 800 BPI (bytes per inch) cassette tape.

3.81 $^{+0}_{-0.05}$ mm
(0.150 $^{+0}_{-0.002}$ in)

Initial gap ②

33 mm Min.
(1.3 in Min.)

Preamble
8 Bits

Data block

Data portion
32 to 2064 bits ③

Postamble
8 Bits

Interblock gap ②

17.8 mm Min.
(0.7 in Min.)

Bot marker

Forward tape motion

Track 2

Track 1

0.37 to
0.51 mm
(0.0146 to
0.020 in)

0.37 to
0.51 mm
(0.0146 to
0.020 in)

1.83 mm Min.
(0.0721 in.)

1.83 mm Min.
(0.0721 in.)

800 BP1 ④

1.45 mm Nom.
(0.057 in)

1. Tape is shown with oxide side out.

② Tape is fully saturated in the erase direction in the interblock gap and the initial gap.

③ The last two characters (16 bits) of the data portion is the Cyclic Redundancy Check (CRC).

④ Shown without phase flux reversals that may exist between data bits.

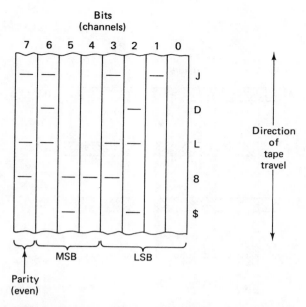

Figure 3-67 Typical tape recording format for ASCII code with eight channels on one track.

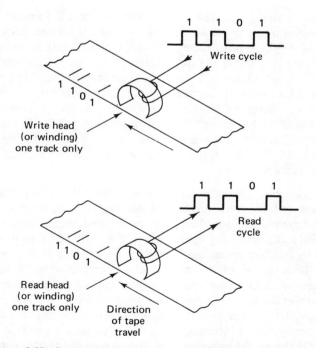

Figure 3-68 Read and write cycles of magnetic tape, showing one bit (on one channel) of four successive data bytes.

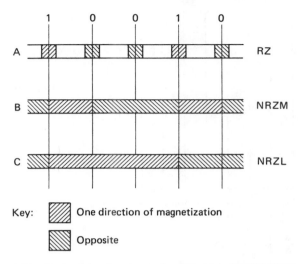

Figure 3-69 Changes in directions of magnetizing for RZ, NRZL, and NRZM data bit recording formats.

most information on a given area of tape. That is, the RZ format provides the lowest amount of *pulse packing* or *packing density*. Pulse packing is a measure of how dense the pulse recording is (how close together the information is recorded). The higher the pulse density, the less tape is needed for storing data. With the RZ method, the pulse current returns to 0 between bits and remains at 0 for a 0 bit. (In some RZ systems, the pulse current returns to 0 between bits, but changes direction for a 1 or a 0 bit.) Pulse density can be increased by use of the NRZL or NRZM formats of Fig. 3-69. In the NRZL format, the direction of magnetization changes only when switching from 0 to a 1, from a 1 to 0. In NRZM, the magnetization changes direction when it is necessary to store a 1.

Tape control systems.   Another difference between home entertainment and digital tape systems is the method of starting and stopping. Unlike an audio tape that moves slowly and need not be stopped, digital tape moves at fast speeds. For a cassette using the format of Fig. 3-66, the tape moves at a speed of 100 in./s when searching for data, and 15 in./s when reading or writing data. Because the microcomputer must pinpoint specific data (at a selected address or block on the tape), the tape control system must be able to stop the tape at the proper point with no lost motion.

Reading and writing on tape is performed on the tape unit at a constant speed. The *transfer rate* of information to and from tape depends largely on two factors: (1) the actual movement of tape across the read/write heads, and (2) the number of characters or data bytes that can be stored on 1 inch of tape (packing density). For example, assume a tape movement speed of 15 in./s and a density of 800 bytes/in. At these rates, the character (or byte) transfer rate is 12,000 per second.

Reel-to-reel tape control.   Several methods are used to move tape across the read/write heads of digital reel-to-reel tape recorders in a fast, yet synchronized manner to prevent breakage. The sudden burst of speed in starting a tape and the abrupt jolt when the tape stops are "softened" by making slack in the tape areas where breakage is likely to occur. Figure 3-70 shows some typical systems.

One system uses vacuum columns to house loops of tape on both sides of the read/write heads. The tape loops are suspended in vacuum columns, and a slight air pressure is applied to the top of the tape to provide some tension. The vacuum columns take up and give the required slack before and after recording. The loop in each vacuum column acts as a buffer to prevent high-speed starts and stops from breaking the tape. Vacuum-actuated switches in the columns allow the *file reel* and *take-up reel* to act independently. The file-reel feeds tape when the loop in the left chamber reaches a minimum point, and the take-up reel winds the tape when the loop in the right chamber reaches a maximum point. During rewinding or backspacing of tape, the two reels simply reverse their roles. The vacuum action is the same, although rewinding speeds generally are faster than reading or writing speeds.

Figure 3-70 also shows a reel-to-reel tape control system using photocell switches. These switches keep the tape loops above the lower switch and below the upper switch. Each tape reel has its own drive motor connected to a servomechanism. When a tape reel starts, it has only to draw from the loop, not from the entire tape. The other reel and its motor drive the tape to keep the loop at the proper size. In this way, quick stops and starts are possible without breaking the tape. In some high-speed systems, the tape reels are fed into bins where the tape lies in folds as shown. The tape is then wound onto reels for future use.

Only the basic tape control systems are shown here. There are many other tape control systems, and new systems are being developed. Most cassette tape systems, and some reel-to-reel systems, have automatic tape threading, where tape cartridges are mounted on the tape drive mechanism for self-threading. Some reel-to-reel tape control units use pinch rollers to move tape. Other units use suction only, with rotating capstans to pull the tape past the read/write heads. Some systems read tape in one direction only; others are capable of reading the reverse direction as well.

Storing data records on tape.   Information is usually written on tape in groups (called *data blocks*), as shown in Fig. 3-66. In normal operation, information continues to be written on the tape until a flag or control signal is

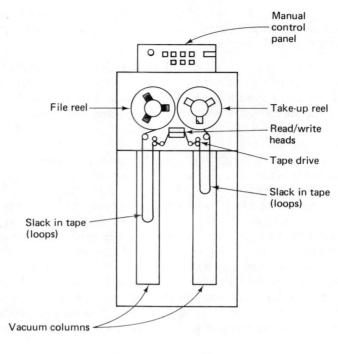

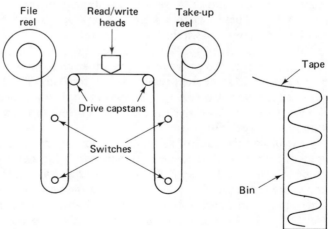

Figure 3-70 Typical control systems for digital reel-to-reel tape recorders.

received from the digital device (such as a computer). This stops the tape drive. Some tape systems use a fixed number of characters in each block, but variable-length blocks are more common. Actual length is determined primarily by the storage capacity of the digital device.

As shown in Fig. 3-66, there are *interblock gaps* between data blocks. (In some systems, the gaps are known as *interrecord gaps,* or IRGs). These interblock gaps are produced when tape continues to be driven, but no information is written on the tape. The use of interblock gaps permits the tape drive to accelerate and decelerate when starting or stopping without failing to read and write the desired information. In some tape systems, when an interblock gap is encountered during the read operation, the tape drive stops at about the center of the gap. The remaining half of a gap is used upon acceleration before the next record is read. A new read command from the digital control circuit starts the tape drive again.

Safety and accuracy of tape records.   Many safeguards must be taken to preserve data stored on digital tapes. Obviously, the information can be destroyed if an erase signal is accidentally applied to the tape unit heads or if new information is written on top of old information.

A common safeguard on large reel-to-reel tape transport systems to prevent accidental overwriting of tape is a safety indicator in the form of a plastic ring. When installed on a groove around the hub of the tape reel, the ring indicates that the tape may be used to store new information. The ring is removed when the writing of new information is to be prevented. This accounts for the term *no-ring, no-write* used by many computer operators and programmers.

The tape transport of a typical cassette recorder contains two microswitches, one to sense the presence of a tape cassette in place, and the other to see if the *write protect tab* (sometimes called the *file protect tab*) is removed. If the cassette tab is removed, the tap is "write-protected" and circuits within the cassette disable the write circuits. Signals from these switches or sensors are available at the interface between cassette and digital device. The digital device must check them prior to issuing any "motion" commands to the tape transport.

Just as important, but perhaps not so obvious, there must be some system to ensure that the data bytes are recorded accurately. There are three main concerns in recording data bytes. First, *all of the bits* must be recorded. Second, the bits must be *readable.* Third, the bits must be recorded in correct order, both horizontally and vertically on the tape. Several systems are used to ensure these conditions.

Dual-recording systems can be used where reliability is particularly important, such as when real-time data bytes are recorded on a one-time-only basis. In dual recording, each data byte is written twice in each frame across the width of the tape by two sets of heads, amplifiers, and so on. Of course, this requires twice the number of channels and tracks. Note that there are two 8-channel tracks on the format of Fig. 3-66.

Some systems provide a *constant check* (for readability) of the data bytes being recorded. These systems are sometimes known as *dual-gap* read/write arrangements. All characters or data bytes written on tape are immediately read by a separate head (or winding) adjacent to the write head. Both readability and accuracy can be checked in this way. A basic constant-check system is shown in Fig. 3-71. In this system, information on a microcomputer output register is amplified and written onto the tape by a write head. Only one bit is shown in Fig. 3-71. The circuit is repeated (amplifier, head, etc.) for each bit in the register. As the tape moves, the bit is read out immediately, amplified, and applied to another readout register. Each bit in the readout register is compared against the corresponding bits in the output register, using a comparator circuit.

The recording operation continues as long as the recorded bits are

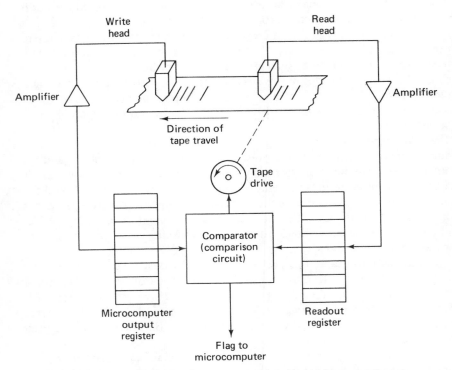

Figure 3-71 Basic constant-check (dual-gap) read/write magnetic tape recording system.

readable and in agreement with the bits in the output register. If there is any inequality between the two registers, the comparison circuit stops the tape drive and sends a flag to the microcomputer. In some systems, the tape drive is reversed, and the bit is rewritten at the appropriate location on the tape.

When dual-recording and constant-check systems are not practical, *parity and CRC* (cyclic redundancy check) *circuits* are used to check the accuracy of bits on the tape. Parity circuits use the parity system (described in Chapter 2) and provide a check of individual characters (data bytes) across the track. Such systems are sometimes known as *vertical check* or *vertical parity* schemes. A basic vertical parity system is shown in Fig. 3-72. In this system, one channel of an eight-channel track is served for vertical even-

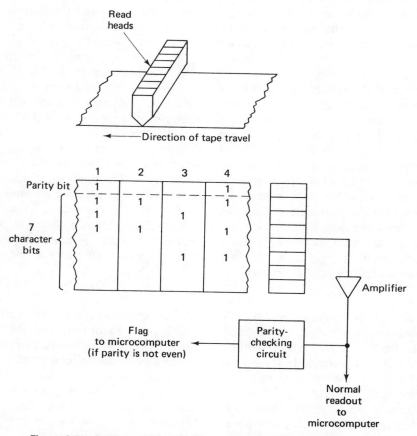

Figure 3-72 Basic vertical parity system for magnetic-tape recording control.

parity check. A 1 bit is added to make an even number of 1 bits across the tape width. In the case of data bytes 2 and 3, both normally use an even number of 1 bits. Thus, no parity bit is added. Data bytes 1 and 4 use an odd number of 1s (three 1s). Thus, a parity bit is added to both data bytes 1 and 4. The amplified output of the read heads applied to the microcomputer is also applied to the parity-checking circuit. If the input to the parity-checking circuit is even, operation continues as normal. If the parity checker receives an odd number of bits (due to recording failure, tape failure, etc.) a flag is sent to the microcomputer.

CRC circuits provide a check of a complete data block, and are thus sometimes referred to as a *horizontal parity* or *horizontal check system.* As shown in Fig. 3-66, the last two bytes or characters of the data portion in the data block are called the CRC bytes. These two CRC bytes are generated by circuits that count all the data bits, starting with bit zero of the first data byte in the block (sometimes known as the address byte) ending with the last bit of the last data byte (excluding the CRC bytes). When the data block is read back from the tape, the data bits (from bit zero of the address or first data byte to the last bit of the second CRC byte) are divided by the CRC number. The remainder should be zero if all the data bits have been properly recorded and played back. A nonzero remainder indicates an error within the data read back. This nonzero remainder can be used to stop the tape drive send a flag to the microprocessor, and so on.

As shown in Fig. 3-66, there are eight *preamble* and *postamble bits* at opposite ends of the data portion of each data block. These are 8-bit patterns or bytes of alternating 1s and 0s. (This is sometimes known as a *checkerboard.*) In some systems, the preamble and postamble bits are used to establish the data rate during data recovery or read operation. In such systems, the tape drive operates at high speeds during gaps, and then slows to a fixed rate when data bytes are recorded or read. The preable and postable bits permit the system to lock onto or synchronize with the desired rate during the read operation.

Locating data on tape records.   Since tape has no physical markings, some means must be used to locate data at a specific point on the tape. This is true for both read and write. In the system of Fig. 3-66, reflective markers (usually silver-coated foil) are attached to the tape at the beginning and end of tape. These are the beginning-of-tape (BOT) and end-of-tape (EOT) markers. Only the BOT marker is shown in Fig. 3-66. A photocell is mounted near the tape and picks up light from the reflective marker. A change in light on the photocell stops the tape drive at the marker until a new read or write command is given to the tape transport.

Although physical markings can be used for the beginning and end of

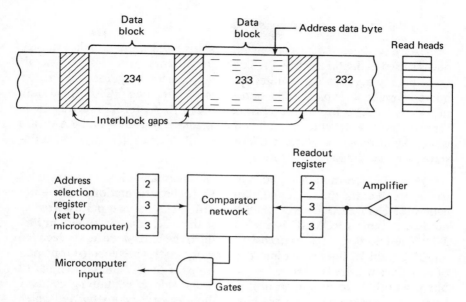

Figure 3-73 Basic circuit for locating data on magnetic-tape records.

a tape, specific information at any point between the markings must be identified by means of an *electronic address*. Several addressing systems are used. In one of the most common, the first character or data byte in the data portion of the data block represents the address for that block. A very simplified version of such a system is shown in Fig. 3-73.

As shown, each data block is identified by a three-digit decimal number (from 000 to 255). The read heads are connected through amplifiers and gates to the microcomputer input. When the gates are opened, the information being read is fed into the microcomputer. The gates are opened upon command from a comparator circuit. This comparator compares the data in address section register with data in the readout register. The address selection register is set to the desired address (data block 233 in the illustration) by the microcomputer, and the tape drive is started at high speed (or *search speed*). When the tape is at the desired address (data block 233), the readout register indicates 233 and the comparator opens the gates. As the tape continues to move, the information in the 233 data block is read into the microcomputer. The tape stops at the end of the block, or goes to a new address, as determined by the address selection register.

Of course, the circuit of Fig. 3-73 is oversimplified for illustration purposes. In practice, the tape drive and address selection circuits move the tape from address to address, stopping, reversing, and again moving forward at high speeds.

3-6-5 Solid-state memories

At one time, ferrite core memories were the most popular storage devices used in digital electronics. Today, ferrite core memories are generally being replaced by solid-state memories, particularly for the smaller microcomputers. Typical solid-state memories include ROMs (read-only memories), RAMs (random-access memories), PLAs (programmable logic arrays), and CAMs (content-addressable memories). ROMs and RAMs are discussed further in Chapter 4. Here, we will describe some typical solid-state memory circuits in IC form.

ROMs. The information stored in a ROM is permanently programmed into the circuit at the time of manufacture. Once the information is entered, it cannot be changed. However, the information can be read out as often as needed. A typical ROM IC is made up of three sections, as shown in Fig. 3-74. These sections are: the *decoder,* in which the binary address is decoded and X-Y pairs of lines going into the memory matrix are enabled (one pair of X-Y lines if there is 1 bit per output; two pairs of X-Y lines if there are 2 bits per output word; and so on); a *memory matrix,* containing as many transistor locations as there are bits in the memory; and a *buffer,* which supplies output levels for the external circuitry.

The physical arrangement of a typical MOS memory matrix containing transistor locations is shown in Fig. 3-75. At the intersection of every X-line and Y-line, any MOS transistor can be either constructed, or omitted, by growing either a thin-gate oxide or a thick-gate oxide. The absence of a MOS transistor is interpreted by the buffer as a logic 0, and the presence of a thin-gate MOS transistor is interpreted as a 1. The programming of the memory (placement of thin-gate oxide transistors) is performed during the manufacturing process.

Aside from orgánization of the ROM, which defines its bit capacity, the most important parameter in most applications is probably *access time* (defined as the time required for a valid output to appear after a valid input has been applied). In a *static ROM,* there are no clocks required. If a valid

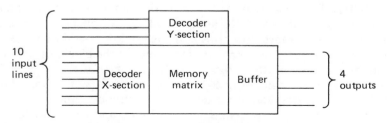

Figure 3-74 Basic sections of a typical IC ROM.

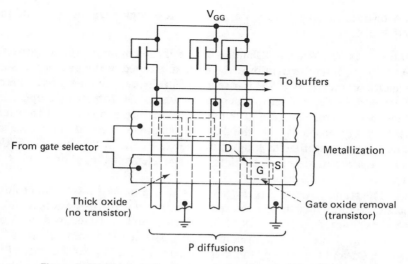

Figure 3-75 Physical arrangement of a typical MOS memory matrix.

input address is applied to the memory, after the expiration of the required access time, a valid output will appear. The output will remain valid as long as the input address remains unchanged. In a conventional *dynamic ROM,* the information is clocked in and clocked out. The output remains valid only for a certain period. Dynamic ROMs are advantageous in implementing synchronous digital circuits. Most dynamic ROMs have latches and buffers to hold the information.

As discussed in Chapter 4, the most common applications for ROMs are to hold the programmed instructions for microcomputers. Other uses for a ROM include operation as a *look-up table,* where the output is a mathematical function of the input; *code conversion* where the input is one code and the output is another code (ASCII for example); *character generator,* where an alphanumeric character is represented by a binary word; and *random logic,* where the ROM is used to perform Boolean algebra (programmed to provide outputs that are Boolean functions of the input variables).

PLA. A PLA is essentially a large ROM with a nonexhaustive decode section adapted to the implementation of random logic. The term "random logic," as applied here, means a logic circuit that is not strongly structured, as opposed to circuits such as shift registers, ROMs, and so on. From a user's standpoint, logic design with PLAs is easy. With this approach, the designer writes down the logic equations of each output in terms of external inputs and feedback inputs. Once this is done, the programming of the PLA matrices is handled by a computer. In effect, from a user's standpoint, the

PLA bridges the gap between a custom MOS logic array and an off-the-shelf ROM.

RAMs. As discussed in Chapter 4, the most common applications for RAMs are to hold the data bytes being manipulated in the microprocessors of microcomputers. A RAM is essentially a fully decoded memory, where a binary address determines the location in which the write or read operation is performed. Both static and dynamic RAMs are available. In the static RAM, an FF is used to store the data, clocks are not needed, and the information stays in storage as long as the power is on. In the dynamic RAM, MOS capacitors are used as data storage elements, and data must be refreshed.

The static RAM cell of Fig. 3-76 is an example of a two-dimensional decode. Assume that the cell is used in a memory with 16 X-address and 16 Y-address lines. Any one of the 256 bits (cells) can be selected by driving one X- and one Y-address line in coincidence. The cell is driven by the X-line through Q5 and Q6, and by the Y-line through Q7 and Q8. The state of the FF (Q1–Q4) is sensed by observing outputs D0 and D1. The same D0 and D1 digit lines are used to write information into an addressed bit.

CAMs. A CAM is a cell in which all the words contained can be matched simultaneously against an argument word, and outputs can be given whenever a *true match* is obtained. Figure 3-77 shows the functions of a CAM, as well as a single CAM cell. Each word has a write input which can also be used to interrogate that word for a match. Two bit lines run through each column of cells, allowing the data word to be written in. The cells may also be used for reading the content of a word or for masking parts of the argument word where an irrelevant (don't care) state exists.

In the basic CAM cell, transistors Q1, Q2, Q9, and Q10 compose an FF for data storage. Transistors Q7 and Q8 are selection transistors which, when turned on by the application of a negative voltage to the W (write) line, connect the FF to the B0 and B1 bit lines. When the W line is at 0, Q7 and Q8 are off, isolating the FF from the bit lines. Transistors Q3 through Q6 and Q11 perform the interrogation logic. For this mode, each W line is grounded through a small resistor and Q11 is turned on. Transistors Q3 through Q6 compare the state of the memory cell FF with the voltages externally applied to the bit lines. Thus, the word lines are controls for write and read operations, but are outputs for the interrogate operation. The bit lines are inputs for write and interrogate, but outputs for read.

3-6-6   Using shift registers as memories

Solid-state shift registers can be used as temporary memories. Two such applications involve use of storage registers as *refresh memories* and *scratch-pad memories*.

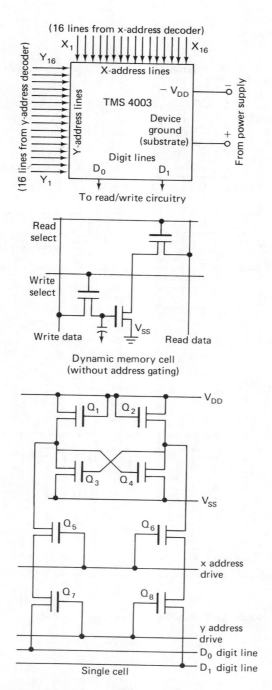

Figure 3-76 Typical static MOS RAM with two-dimensional decoding.

209

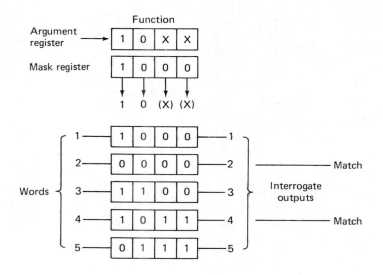

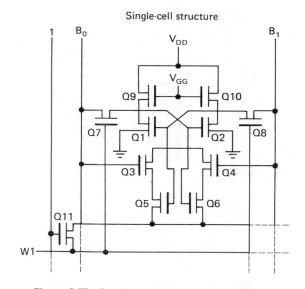

Figure 3-77 Functions and structure of a CAM.

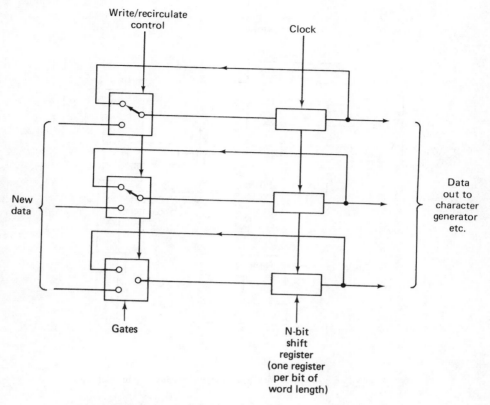

Figure 3-78 Using an N-bit shift register as a refresh memory.

Figure 3-78 shows an N-bit shift register used as a refresh memory by returning output to inputs. Such an arrangement is particularly useful for renewing fading displays such as CRT character-generator systems (discussed in Chapter 4). New information is written via a two-way input gate circuit as shown. The rate (in seconds) at which a particular bit of information is available at the output is determined by the expression "N/clock frequency."

By adding an address counter and comparator, the refresh memory becomes a scratch-pad memory, as shown in Fig. 3-79. Information can be written into and read out of any point specified by the input address code. An output register is necessary to store the required output data and to provide a 1-bit delay so that the read address is the same as the write address (because there is a 1-bit delay between output and input).

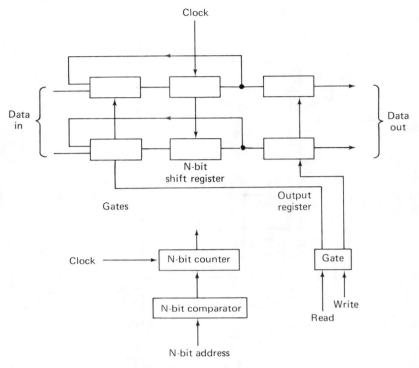

Figure 3-79   Using an N-bit register as a scratch-pad memory.

3-6-7   Other memories and storage units

Thus far, we have discussed only the most popular memories and storage units found in digital electronics. (Chapter 4 also discusses the practical details of ROMs, RAMs, and floppy disk memories.) There are many more types of memories in existance, and new memory systems are being developed constantly (particularly solid-state memories). We do not discuss these memories because of the space limitations. (An entire book could be devoted to digital electronic memories). For this reason, we have concentrated on how memories are integrated into digital electronic systems. Once you understand these practical basics, you will have no difficulty in understanding operation of new memory systems as they are developed. Some typical digital electronic memories include magnetic drum, nonflexible disk, magnetic card, optical disk, magnetic bubble, and charge-coupled devices (CCDs).

CHAPTER

4

Typical Digital Computer Systems

This chapter describes operation and characteristics of typical digital computer systems. Since microcomputers are currently the most popular form of digital computer, we will concentrate on the microcomputer. Because of the complex nature of even the simplest computer, we will not attempt to cover the full details in this one chapter. Instead, we will concentrate on microcomputer basics. For a thorough analysis of present-day computers, your attention is invited to the author's best selling *Handbook of Microprocessors, Microcomputers, and Minicomputers* (Englewood Cliffs, N.J.: Prentice-Hall, Inc., 1979).

4-1   MICROCOMPUTER TERMS

The heart of a microcomputer is the *microprocessor*. The microprocessor is an IC that performs many of the functions found in a digital computer. A single microprocessor IC is capable of performing all of the arithmetic and control functions of a computer. By itself, a typical microprocessor IC does not contain the memories and input/output (I/O) functions of a computer.

However, when these functions are provided by additional ICs, a microcomputer is formed.

Typically, a basic microcomputer requires a ROM to store the computer program or instruction, a RAM to store temporary data (the information to be acted upon by the computer program), and an I/O IC to make the system compatible with outside (or peripheral) equipment such as an interactive video computer terminal, teletype, or line printer. There are some ICs that contain some, or all, of these functions. In effect, when an IC contains all of the functions, the IC is a "computer on a chip." However, this is not the typical case.

Microprocessors are sometimes referred to as *microprocessor units* (MPUs) or *control processor units* (CPU) (CPU can sometimes mean a *central processor unit*. It should also be noted that a microprocessor is not always used in digital computer applications. Instead, the microprocessor is used as a controller. As a matter of interest, the microprocessor was originally developed as the control element for those applications where digital computer function (the ability to store and execute a complete program automatically) is required, but where a computer (even a simple minicomputer) is too large or expensive. Sometimes, the microprocessor is called a *microcontroller* when used in these control applications. The various terms are discussed further throughout this chapter.

The term "minicomputer" can be applied to many relatively small and relatively simple computers. A minicomputer often contains many ICs, but not necessarily a microprocessor IC. In this chapter, we are concerned primarily with microprocessor-based systems, whether they be computers, controllers, or whatever.

4-2   THE ELEMENTS OF A MICROCOMPUTER

As discussed in Chapter 1 (and illustrated in Fig. 1-4), a microprocessor performs its function (program counting, addition, subtraction, etc.) in response to instructions in the form of electrical pulses (arranged to form a binary word). For one microprocessor, this word may be an instruction to perform addition. For another microprocessor, the same word may be meaningless. Thus, one of the first things you must do to understand and use a microprocessor is to learn all its instructions (known as the *instruction set*), including the corresponding binary code, hex code, and what is accomplished by the microprocessor when the instruction is received. Microprocessor manufacturers provide information on the instruction set in their manuals. This is fortunate, since it is nearly an impossible task to remember the entire instruction set of even a simple microprocessor.

4-2-1   Microprocessor memories and addresses

Microprocessors also use electrical signals arranged in binary form to communicate with other elements in the systems. Most microprocessors are used with memories (RAMs or ROMs, typically both) which hold *data bytes* to be manipulated by the microprocessor, and *instruction bytes* to be followed by the microprocessor during the program. (Data bytes are usually stored or held in the RAM; instructions are usually stored in the ROM.)

Memories are divided into locations called *addresses*. Each address is identified by a number (usually decimal). During a typical microcomputer program, the microprocessor selects each address in a certain order (determined by the program) and reads the contents of the address. Such contents can be an instruction, data, or a combination of both. Similarly it is possible to write information into memory using electrical signals arranged in binary form.

4-2-2   Buses and ports

A typical microprocessor/memory arrangement is shown in Fig. 4-1, where a microprocessor is connected to a RAM and a ROM by data and address *buses*. The term *bus* is applied when several electrical lines are used for a common purpose. In our example, the data bus has eight lines (which is typical) and the address bus has eight lines (which is not typical). Generally, the address bus will have as many as 16 lines. The eight-line system is used here for simplicity. The term *highway* is sometimes used when a bus is used to interconnect many system components. Also, the term *handshake bus* is sometimes used to indicate a bus that interconnects a microprocessor system with the outside world.

The term *port* is often applied to the point or terminals at which the bus enters the microprocessor or other IC element. Thus, in the microprocessor of Fig. 4-1, there is an address port and a data port. Buses are generally bidirectional. That is, the electrical pulses (representing data bytes, addresses, etc.) can pass in either direction along the bus lines. For example, data bytes can be written into memory from the microprocessor, or read from memory into the microprocessor on the same data bus. Ports may or may not be bidirectional, depending upon the design.

4-2-3   Address and data buses

Note that in Fig. 4-1, the electrical pulses appearing on the address bus are arranged to produce a binary 01001101, or a decimal 77. Thus, address number 77 is being selected by the microprocessor. Both the ROM and RAM receive the same set of electrical pulses (or binary word), since these

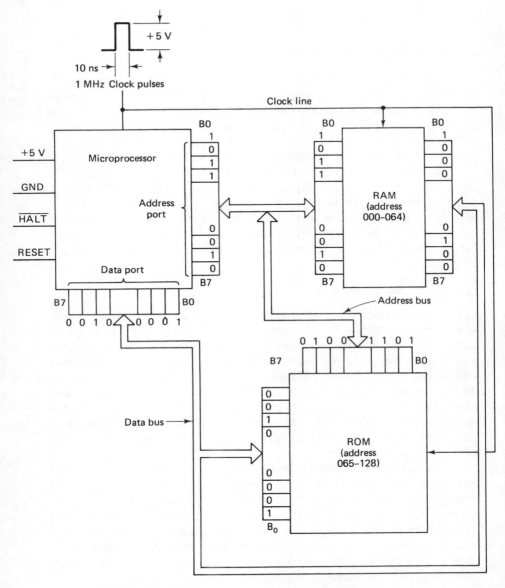

Figure 4-1 Typical microprocessor memory arrangement, including data and address buses.

pulses appear on the address bus. However, since address 77 is located in the ROM, the contents of the ROM at that address are read back to the microprocessor via the data bus. The pulses on the address bus have no effect on the RAM, and no information is obtained from the RAM. Similarly, the address pulses have no effect on other addresses in the ROM. Only the information at the selected addresses is read back on the data bus. In Fig. 4-1, the electrical pulses on the data bus are arranged to form the binary word 00100001, which can be converted to hex 21. In our particular microprocessor, hex 21 is an instruction to add the contents of a register within the microprocessor to the contents at some address in the RAM (read out during a previous step in the program).

4-2-4 Debugging and troubleshooting

Keep in mind that the microprocessor responds to the *arrangement* of electrical pulses. It makes no difference where the pulses originate. Thus, if an undesired instruction is stored in memory at some location which is addressed by the microprocessor during a program, that instruction will appear as the corresponding arrangement of electrical pulses on the data bus. The microprocessor will follow the instruction when received and will thus produce an erroneous result. The program will probably come to a complete halt or jump to an undesired address. Under these conditions, the program is said to have a *bug* or *bugs.* The process of finding the undesired instruction, removing it, or placing it at the correct address is known as *debugging,* which applies to finding any fault in a program.

This is not to be confused with *troubleshooting,* which is the term used to find electrical or mechanical faults in a microprocessor system. Troubleshooting implies that the system once performed the program properly. For example, referring to Fig. 4-1, assume that the electrical line labeled BO on the data bus becomes broken (after weeks of operation) at the microprocessor terminal. The BO pulse from the ROM still appears on the data bus but does not reach the microprocessor. Thus, the microprocessor sees zero volts, or binary 0, on terminal BO. This produces binary word 00100000, or hex 20, instead of the desired binary 00100001, or hex 21. The hex 20 might be meaningless or might be an erroneous instruction. In any event, the microprocessor will not perform the correct function.

4-2-5 Nature of pulses

The pulses used by microprocessors are instantaneous electrical signals. Typically, pulses start at zero volts, rise to 5 V, and then drop back to 0 V, all within a few nS or μS. All the pulses in a given binary word or data byte must arrive at the same time. Thus, all the pulses are said to be transmitted in *parallel* on the bus. All binary information within a microprocessor system is transmitted in parallel form.

This is not necessarily true in the world outside the microprocessor system, where information can also be transmitted in *serial* form. In serial transmission of an 8-bit binary word, eight pulses (for binary 1) or spaces (for binary 0) are transmitted at regular intervals on a single electrical line (or pair of lines). This is followed by a long space before the next 8-bit byte is transmitted. Obviously, serial is slower but requires only one or two lines, compared to one line for each bit in parallel. Since microprocessors use only parallel within the system, serial data bytes must be converted before they are used with a microprocessor system (and vice versa). This is one of the functions of an I/O device.

4-2-6   Clock and timer pulses

The electrical pulses used by the microprocessor, ROM, and RAM are generated by a *clock* or *timer,* which may be part of the microprocessor or can be external. In Fig. 4-1, the clock is external and is a +5-V pulse of 10-ns duration, at a frequency of 1 MHz. The clock circuit is an oscillator that produces pulses of fixed amplitude and duration at regular intervals. One to 3 million pulses per second (or 1 to 3 MHz) is a typical clock frequency.

The microprocessor, ROM, and RAM do not actually generate the binary pulses but produce their binary word pulses on the address and data lines when they receive clock pulses. It may take many clock pulses to form a binary word. In some cases, the microprocessor, ROM, and RAM must also receive other signals before they will produce the binary word pulses or data byte. For example, a RAM usually requires a "read" signal, plus the clock signal, before the contents of an address is read onto the data bus from the RAM. Such control signals are discussed in later paragraphs of this chapter.

4-2-7   Power supply and other signals

Not all electrical voltages and signals applied to a microprocessor, ROM, or RAM are in pulse form. For example, referring to Fig. 4-1, note that there are two lines into the microprocessor, labeled +5 V and GND. These are the power supply lines connected to an external 5-V power supply.

There are also two lines labeled $\overline{\text{HALT}}$ and RESET, respectively. The RESET and $\overline{\text{HALT}}$ lines receive a +5-V signal from various circuits in the system. This signal may be a momentary pulse identical to those on the address and data buses, or may be a fixed +5 V which remains on the line for some time. When a fixed signal (sometimes called a *level*) is applied, the line is said to be "high" and the function is "turned on."

For example, if +5 V is applied to the RESET line, the RESET line is high and all circuits within the microprocessor are reset to zero, regardless of their condition before the line goes high. When the fixed voltage is removed, the RESET line goes low (is at zero volts) and the RESET function is no longer in effect. In most microprocessors, when a reset signal is received, all circuits return to zero and then resume their normal function (counting, etc.).

An overbar is used on the word $\overline{\text{HALT}}$. This indicates that the $\overline{\text{HALT}}$ operates on the reverse of all other lines. That is, the $\overline{\text{HALT}}$ function is in effect when the line is at 0 V (the normal low condition). When the line is at +5 V (normal high), the $\overline{\text{HALT}}$ function is removed. In our microprocessor, when the $\overline{\text{HALT}}$ line is at 0 V, all functions within the microprocessor (counting through the program, etc.) are stopped and remain stopped as long as the $\overline{\text{HALT}}$ line remains low. All functions resume normal operation when the $\overline{\text{HALT}}$ line is made high by a + 5-V level. If the microprocessor is in the middle of some operation when $\overline{\text{HALT}}$ is applied (by 0 V on the line), the operation will stop, but it will continue from the same point when the + 5 V is reapplied.

This illustrates the need to understand *all instructions and control signals* applied to a particular microprocessor. A thorough knowledge of microprocessor functions and controls is essential for writing and debugging programs as well as troubleshooting. For example, should the program inadvertently issue an instruction that removes the + 5 V from the $\overline{\text{HALT}}$ line, the microprocessor will stop in the middle of a program, possibly in the middle of an instruction. The same condition can be caused if the $\overline{\text{HALT}}$ line is accidently disconnected from the $\overline{\text{HALT}}$ terminal on the microprocessor, if the $\overline{\text{HALT}}$ line is shorted to ground, or if the $\overline{\text{HALT}}$ line is broken.

4-3 MICROCOMPUTER CIRCUIT FUNCTIONS

All circuit elements of a microcomputer system are in IC form. Possible exceptions are the interconnecting buses, wiring, and a very few external switches, gates, readouts, keyboards, and so on. In any event, the microprocessor, ROM, RAM, and I/O device are all IC. Thus, you do not have access to the circuit elements, nor can you change them in the way they function. Also, it is not necessary that you understand every detail of the internal circuits to effectively use and understand microprocessors. In Chapters 2 and 3, we discussed the operation of detailed circuits such as registers, gates, and ALUs. Here, we review operation of these circuits as they relate to microprocessor-based devices.

4-3-1 Counters, registers, accumulators, and pointers

For our purposes here, counters, registers, accumulators, and pointers are all circuits used to hold and manipulate binary numbers in electrical (level or pulse) form. As such, these circuits have one stage for each binary bit to be held or manipulated. Thus, an 8-bit counter/register has eight stages. As discussed in Chapter 2, FF stages are used in counters/registers, and an FF can be in only one of two electrical states, 1 or 0. If you measured the instantaneous state of a particular stage, you would find it at + 5 V if the stage is to represent a 1, and at 0 V for binary 0. (Except in certain troubleshooting situations, it is not necessary to actually measure the states of counter/register stages. However, this concept may help you to understand operation of counter/registers.)

In a microprocessor, the purpose of a counter is to count events (such as steps of a program, a sequence of addresses selected, etc.). This is usually done by counting pulses. Counters are sometimes used as *pointers,* in that they point to another event or location. For example, a typical program counter counts each step of the program and then advances to the next address to be used in the program. Thus, the counter "points" to the next step of the program. Note that when a counter, register, or another circuit in a microprocessor system is used solely or primarily for one purpose, it is said to be *dedicated.* For example, a microprocessor counter used only to count program steps is referred to as a *dedicated counter.*

A microprocessor register is similar to a counter except that the primary function of a register is to hold the binary numbers (or words) so that they can be manipulated. Registers are often used to hold some binary number taken from a particular address in memory so that number may be added to another number in memory. Similarly, one register can hold a binary number that is to be added to another binary number in another register. When a register is used primarily for arithmetic operations, the word *accumulator* is often applied.

Serial operation. Microprocessor literature uses many methods or symbols to represent counters and registers. One of the most common methods is shown in Fig. 4-2, which illustrates operation of a typical counter used to count serial pulses. Assume that the circuit is used as a program counter and that it receives one pulse for each step of the program. That is, the counter is to be *incremented* (or advanced) for each pulse representing a step. (When each step or pulse removes one count, the counter is said to be *decremented.*)

Initially, all stages are *reset* to *cleared* to low, binary 0, or 0 V, by a reset signal. In serial operation, the first or LSB stage operates on the pulses to be counted. All other stages receive a pulse from the stage ahead. As each pulse to be counted is applied, the stage representing the LSB is changed from a 0 to 1 (0 V to +5 V). When the next pulse arrives, the first stage returns back to 0 V (binary 0) and sends a signal to the next stage, which moves to binary 1. This process continues until all pulses are counted. If there are four pulses, as shown in Fig. 4-2, the third stage moves to 1 and the first two stages are at 0. All remaining stages are at 0. The counter indicates a binary 00000100 (decimal 4) and corresponds to the number of pulses applied and, in turn, to the number of steps in the program accomplished thus far. In this way the instantaneous count corresponds to the

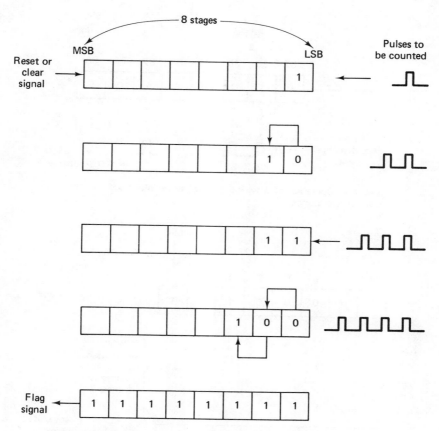

Figure 4-2 Operation of a typical counter or register used to count serial pulses.

program step just accomplished, or to be accomplished next, depending upon design.

When all stages are moved by sufficient pulses to 1 (the binary count is 11111111, decimal 127), the counter is full. Generally, the counter sends a flag or signal to other circuits, indicating the full count. This flag can be used to halt or reset operation of the microprocessor if the program had only 127 steps. Or, the flag can be used as the first pulse to the LSB of another counter (to accommodate a 16-bit word). It should be noted that flag signals do not always indicate a full count in microcomputers. The term "flag" can be applied to any signal which indicates that a particular condition has occurred (full count, error, request for further information, etc.).

Parallel operation. Counters and registers can also be designed to receive information in parallel form, as shown in Fig. 4-3. Here, all the stages are

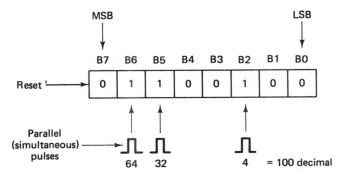

Figure 4-3 Operation of a typical counter or register used to count (or receive) parallel pulses.

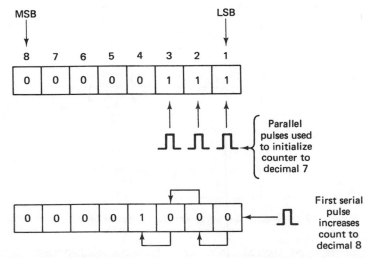

Figure 4-4 Operation of a typical counter or register initialized to decimal 7 by parallel pulses, and increased to decimal 8 by first serial pulse.

set simultaneously by pulses in the binary word 01100100 (decimal 100). That is, bits 2, 5, and 6 receive +5-V pulses, and all other bits remain at 0 V. Sometimes, the terms *load* or *dump* are used when data bytes are so applied to a register. Typically, a register will hold the data byte (all stages remain at this selected state) until the register is cleared (reset to zero) or until another set of pulses that form a data byte is applied (or the power is removed).

Initializing. This ability to be set by both serial and parallel pulses makes it possible to *initialize* counters and registers. It is not always desired to start all counts from zero, or that all counts go through every step in the count. For example, assume that the count is to start at the seventh step of a program. This could be done by applying +5-V pulses to bits 1, 2, and 3 simultaneously, as shown in Fig. 4-4. Then the first serial pulse to be counted moves bit 1 to 0, which in turn moves bit 2 to 0, bit 3 to 0, and bit 4 to 1.

Shifting. The contents of counters and registers can also be shifted by an appropriate signal. That is, the contents of each stage in the counter or register are shifted by one position to the right or left by a shift signal applied to all stages simultaneously. The effects of a left shift are shown in Fig. 4-5, where a register is holding the binary word 00001000 (decimal 8). Bit 3 is at binary 1, and all remaining bits are at 0. During the shift, the contents of bit 0 move to bit 1, bit 1 to bit 2, and so on. After the shift, bit 0 remains at binary 0, since there is no new pulse entering bit 0. Bit 4 is at binary 1, since this was the state of bit 3 before the shift. All other bits are at 0, since 0 was the state of corresponding stages to the right (before the shift).

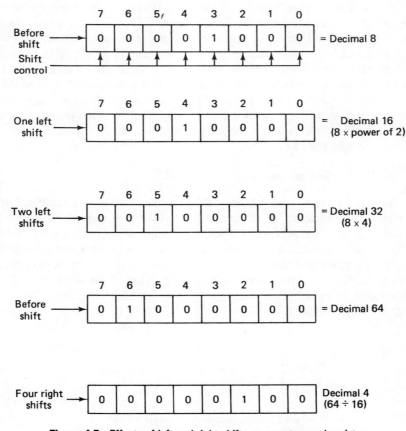

Figure 4-5 Effects of left and right shifts on counters and registers.

As a result of the shift, the binary number is changed from 00001000 to 00010000 or from decimal 8 to decimal 16. Thus, one left shift multiplies the number by the power of 2. If there are two shifts to the left, the binary number is changed to 00100000, decimal 32 (or the same as multiplying by 4). Three left shifts produce 01000000, decimal 64 (multiplication by 8). Registers can also be shifted to the right, which results in division by the powers of 2. This is also shown in Fig. 4-5, where a register is shifted four places to the right for a division by 16. That is, binary 01000000 (decimal 64) is shift to binary 00000100 (decimal 4); 64 divided by 16 is 4.

4-3-2 Decoders and multiplexers

The terms "decoder" and "multiplexer" are often interchanged in microprocessor literature. As discussed in Chapter 2, in a strict sense, a decoder converts from one code or numbering system to another, such as from BCD to seven-segment, and so on. Equally strict, a multiplexer (or MUX) is a data selector and/or distributor. However, in microprocessor literature, the terms are generally applied to any circuit that converts data from one form to another. For example, one microprocessor contains a circuit (designated by the manufacturer as a multiplexer) that converts 16 lines of information from a register into eight lines suitable for an eight-line address bus. Another microcomputer system contains a decoder in each ROM, which makes it possible to select one of 128 memory addresses with an 8-bit word supplied on an 8-bit address bus. The symbol for such decoders and multiplexers is usually a box with the appropriate number of lines in and out, as shown in Fig. 4-6.

4-3-3 Buffers, drivers, and latches

Buffers, drivers, and latches are also terms often interchanged in microprocessor literature. A buffer can be considered as any circuit between two other circuits that serves to isolate the circuits under certain conditions and to connect the circuits under other conditions. A buffer circuit between an internal register and the data bus (at the data port) is a classic example, as shown in Fig. 4-7. Here, the buffer can be switched on or off by a *bus enable* signal or pulse. (This same signal can also be known as an *enable, select,* or *strobe* signal.)

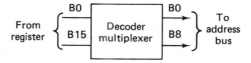

Figure 4-6 Basic decoder/multiplexer symbol.

When the enable signal is present, the buffer will pass data, instructions, or whatever combination is on the data bus to the register within the microprocessor. When the enable signal is removed, the buffer will prevent passage of data. This feature is necessary, for example, when the register is holding old data not yet processed, but there are new data bytes on the bus. The buffer closes the data port until the register is ready to accept new data. This raises the obvious problem of what happens when the new data byte is momentary. In such cases, the buffer has a "latch" function (usually an FF for each bit) which permits each bit in the buffer to be latched to a 1 or a 0 by the data pulses.

Three-state buffers. Typically, buffers are three-state devices, with one stage for each bit to be handled. The stages can be in one of three states: *in, out,* or at a *high-impedance* level. This high-impedance level or state makes it appear that the circuit is closed to the passage of data. Some buffers also include a *driver* function, particularly where the buffer is going to feed

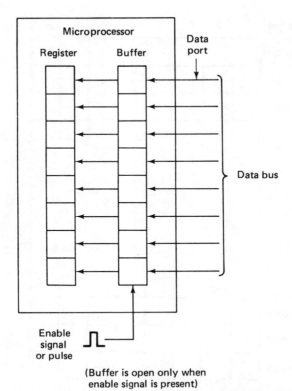

(Buffer is open only when enable signal is present)

Figure 4-7 Buffer circuit between an internal register and the data bus.

many devices simultaneously via a bus. The output of a register or counter may be sufficient to drive one other register, but not many registers or devices. Thus, the buffer includes a driver function which amplifies the drive capability.

4-3-4 Matrix

The term matrix is generally applied to the circuits within the RAM, ROM, or ferrite core memory, but is also applied by some manufacturers to a group of registers in the microprocessor. A typical memory matrix is divided into sections or location, with each section identified by an address, as shown in Fig. 4-8. In turn, the addresses are divided into a number of stages, with one stage for each bit to be held in memory. Thus, each address in an 8-bit memory has 8 bits. The number of addresses in a memory matrix depends on use. Generally, the number is based on the powers of 2, since the addresses are selected by a number system based on binary (such as hex).

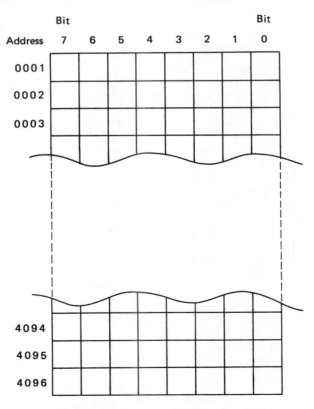

Figure 4-8 Typical 4096 X 8 memory matrix.

A typical ROM matrix will have 128 addresses, each with 8 bits, and will be described as a 128 × 8 matrix. A typical RAM matrix can have 512, 1024, or 4096 addresses (or possibly more). As discussed in Sec. 3-6, a 1024 × 8 memory might be described as a 1K memory since there are *approximately* 1000 addresses available. Similarly, a 4096 × 8 memory can be called a 4K memory.

Matrix vectors.   Also as discussed in Sec. 3-6, the addresses within a matrix are selected by some form of *vector* or *intersect* system, such as that shown in Fig. 4-9. The memory address to be read in or read out is selected when a

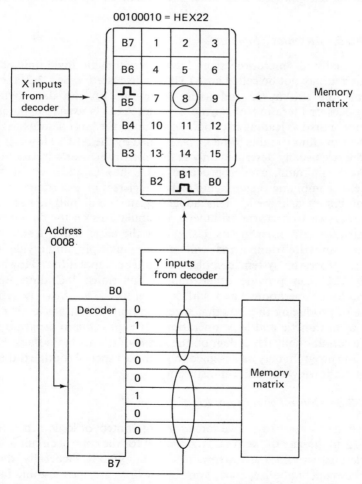

Figure 4-9 Vector system for selecting one address in a 15-address memory matrix.

combination of two appropriate bit lines have a +5-V pulse present (are at binary 1). Only one of the X lines and one of the Y lines can be binary 1 at any given time. All other lines are at binary 0. This action is controlled by some form of decoder or multiplexer. As an example, assume that address 8 is to be selected. The desired decimal 8 is converted to hex 22 (00100010) by a decoder/multiplexer. This causes bits B1 and B5 to receive pulses (+5 V or binary 1). All remaining lines are at binary 0 (zero volts). Thus, the address (number 8) at the intersection or vector of the two lines is selected. Keep in mind that there are 8 bits in each address, and that each of the 8 bits is fed back to the microprocessor or other destination on a bus (usually the data bus).

4-3-5 Arithmetic logic unit

Virtually all microprocessors have an arithmetic logic unit or ALU (although it may not be called an ALU). As discussed in Sec. 3-2, the ALU performs the arithmetic and logic operations on the data bytes. In microprocessor literature, the symbol for an ALU is usually a simple box with lines or arrows leading in or out. Sometimes, the box contains some hint as to the functions capable of being performed by the ALU. However, these functions are usually described only in the microprocessor's instruction set.

As a minimum, a microprocessor ALU has an adder circuit that is capable of combining the contents of two registers in accordance with the logic of binary arithmetic. This adder/register combination permits the microprocessor to perform arithmetic manipulations on the data obtained from memory and other inputs. Using only the basic adder, a skilled programmer can write routines that will subtract, multiply, and divide, giving the microprocessor system complete arithmetic capabilities. However, a typical ALU can provide other built-in functions, including Boolean algebra, logic operations, and shift capabilities. The ALU is generally capable of producing flag bits that specify certain conditions arising in the course of arithmetic and logic operations. It is possible to program *jumps* that are conditionally dependent on the status of one or more flags. For example, the program may be designed to jump to a special routine if the carry bit is set following an addition instruction.

4-3-6 Control circuitry or logic

All microprocessors have some form of control or logic circuit (or circuits). As in the case of the ALU, the symbol for the control circuit is usually a simple box with lines and arrows leading in and out. Generally, the halt, reset, interrupt, initialize, start, and similar lines are shown going into the control logic box symbol. This is because the control logic is the primary functional circuit or unit within the microprocessor. Using clock inputs, the control logic maintains the proper sequence of events required for any processing task.

For example, after a microprocessor instruction is taken from memory (or "fetched") and decoded, the control logic issues the appropriate signals (to the microprocessor, and to such external units as the ROM, RAM, and I/O) for initiating the proper processing action (such as a "write" signal to write data into memory, a "read" signal to read data from memory, and so on).

One function found in most microprocessor control logic circuits is the capability of responding to an *interrupt* signal or *service request* (say from an external video terminal, disk, tape reader, etc.). An interrupt request causes the control logic to interrupt the program execution temporarily, jump to a special routine to service the interrupting device, and then automatically return to the main program.

4-4 MICROCOMPUTER HARDWARE

The term *hardware* applied to a microcomputer system refers to the physical components, wiring, and so on. This contrasts with the term *software,* which applies to programs, instructions, and the like. Some microprocessor manufacturers also use the term *firmware* to describe something between hardware and software. For example, when instructions (software) are permanently programmed into a ROM (hardware), the result is firmware.

The microprocessor of one manufacturer will have little in common with the microprocessor of another manufacturer, with the possible exception of outward physical appearance. Compare the two microprocessors shown in Fig. 4-10. Both are 40-lead, dual-in-line (DIP) IC packages. Both

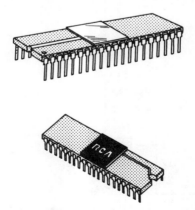

Figure 4-10 RCA and Motorola microprocessors in 40-lead dual-in-line (DIP) IC packages.

are about 2 inches long and ½ inch wide. Now compare this to the terminal assignment diagrams of Fig. 4-11. Although both have 40 leads (also known as *pins* or *terminals*), the terminals are used for entirely different purposes.

For example, the RCA microprocessor has eight data lines to a data bus and eight memory lines to an address bus, whereas the Motorola unit has eight data lines (D) and 16 address lines (PA and PB). From this, it can be seen that you must have all available data on a particular microprocessor to use the unit effectively. Fortunately, such information is available, although the format used to show the interval arrangement of microprocessors will vary from manufacturer to manufacturer.

4-4-1 Microprocessor architecture

The term *architecture* is most accurately applied to the arrangement of counters, registers, ALUs, and so on, within the microprocessor. However,

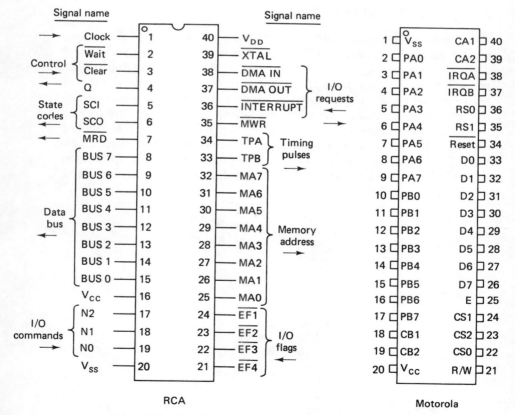

Figure 4-11 Typical microprocessor terminal assignment diagrams.

some manufacturers apply the term "architecture" to the entire system arrangement. There are two commonly used methods to show architecture. The *block diagram* of Fig. 4-12 shows all the internal registers, counters, and so on, of the Motorola M6800 microprocessor of Figs. 4-10 and 4-11. Com-

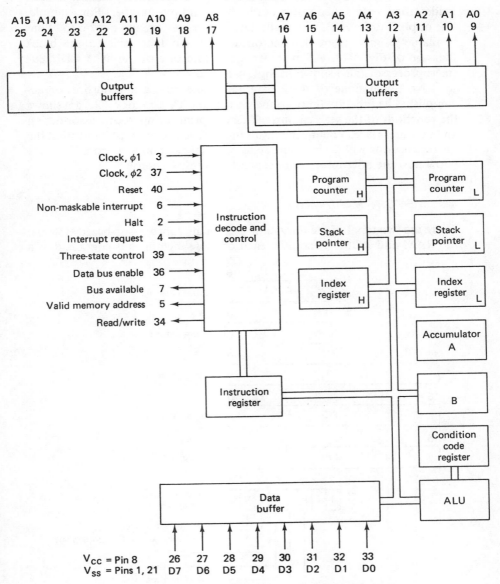

Figure 4-12 Motorola M6800 microprocessor, block diagram.

pare this to the *program model* of Fig. 4-13, which shows only accumulator, register, counter, and pointers. Obviously, the information shown in Fig. 4-13 is suitable for programming only. The information in Fig. 4-12 is required for design and service, since each line to and from the microprocessor is identified as to function (or destination) and pin number. For example, line or pin 2 is the HALT line into the microprocessor, and pin 7 is the "bus available" signal from the microprocessor. Compare this to the block diagram of Fig. 4-14, which shows the architecture of the RCA CDP1802 microprocessor (also shown in Figs. 4-10 and 4-11).

Note that none of these diagrams show how a microprocessor accomplishes its function (in the same sense that a TV set block diagram shows the functions of the set's circuits). This is typical for microprocessor design and user literature. You must consult the instruction set to find out what the microprocessor will do in response to commands. You will probably never know how the microprocessor accomplishes this response.

4-5 MEMORY HARDWARE

There are two types of memory units used in microprocessor-based systems; the RAM and ROM. Generally, each of these memories is contained in a

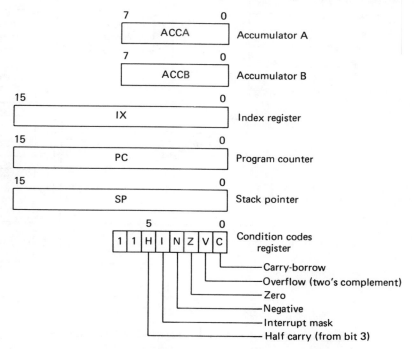

Figure 4-13 Motorola M6800 microprocessor, programming model.

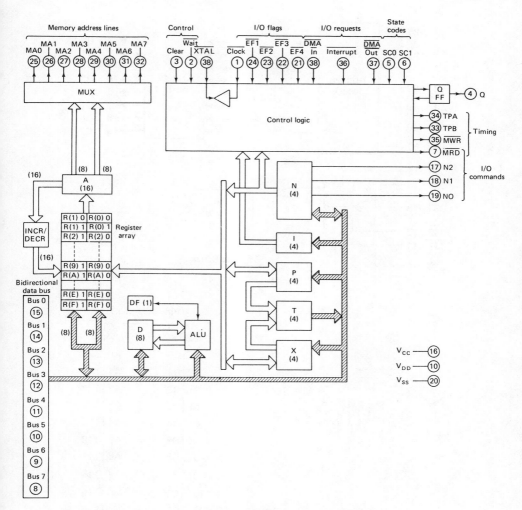

Figure 4-14 RCA CDP1802 microprocessor block diagram.

separate IC package. However, there are systems where one of the memories, or both, are contained in the same IC package as the microprocessor (such as the Intel MCS-48 microprocessor, known as a "computer on a chip").

4-5-1 RAM circuits

Figure 4-15 shows the package, pin assignment diagram, and functional block diagram of a typical RAM. This unit is described by the manufacturer as a 128 × 8 static RAM, meaning that the maxtrix has 128

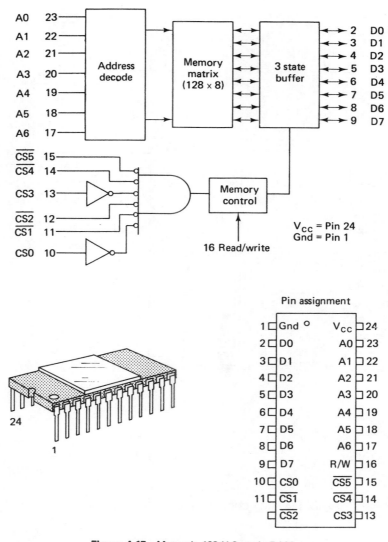

Figure 4-15 Motorola 128 X 8 static RAM.

addresses, each with 8 bits. Thus, 8-bit bytes can be read into, or out of, 128 locations. The data bytes appear on the data lines D0 through D7, usually connected to a data bus. The data byte is read into the matrix if the buffer is in the "in" state, and is read out when the buffer is in the "out" state. The data lines are disconnected from the matrix if the buffer is in the "high-impedance" state.

Buffer-state control. The buffer state is controlled by two factors; the read/write signal at pin 16, and the control signals at pins 10 through 15. If any one of the control signals (called "chip select inputs") is absent, the buffer remains in the high-impedance state, and no data bytes pass between the matrix and the data lines. Thus, it is possible to shut off the memory from the microprocessor system by controlling only one of six lines or inputs.

Note that some of the chip select inputs are *active high* (turned on when the line is at binary 1), whereas others are *active low* (turned on when the line is at binary 0). Active lows are indicated by the overbars. This arrangement permits one memory IC to be turned on, with other memories to be turned off, by the same signal on the same address line. When all the chip select inputs are present (all active highs are at 1; all active lows are at 0), the buffer can be placed in the read (data into the matrix) or write (data from the matrix to the data line) conditions by a signal on the read/write line.

Address selection. The address to be read in or out is selected by an address decoder, which is, in turn, controlled by the binary word on the address bus (lines A0 through A6). In certain systems, part of the address lines are connected to some of the chip select inputs. This is done where there may be several memories, all connected to the address and data buses. For example, assume that a 128 × 8 RAM is used on the same data and address buses with a 1024 × 8 ROM, as shown in Fig. 4-16. Usually, the RAM is assigned to

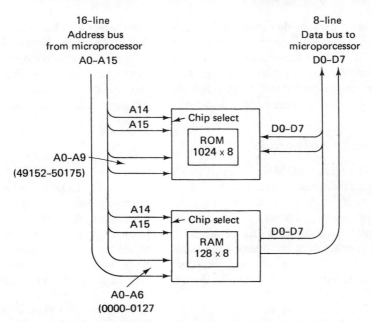

Figure 4-16 Typical address and data bus arrangement where one address line (A15) is used for chip select.

lower number addresses (say from 0000 to 0127, in decimal) with the ROM containing the higher-number addresses (say from 49152 to 50175, in decimal). When the highest address line A15 is at 1, the ROM is made active; when A15 is at 0, the RAM is active. Keep in mind that the use of address line signals control memory ICs is not used in all systems. There are many other arrangements.

4-5-2   Static versus dynamic RAMs

In a typical dynamic RAM, information is stored as an electrical charge on the gate capacitance of a MOS transistor. Since MOS transistors have some leakage, the charge is removed in time, even with the power applied. Thus, the information (binary 0 for 1 for each bit at each address) is lost in time if the charge is not periodically refreshed. As discussed in Sec. 3-6-6, there are many ways to refresh a dynamic RAM. The circuitry is complex and usually involves circuits outside the RAM IC. The main advantages of a dynamic RAM are high-speed operation and low power consumption.

In a typical state RAM, each bit of information is stored on a flip-flop or latch. Static RAMs do not require refreshing and are thus far less complex than dynamic RAMs of the same capacity. However, static RAMs are slower and consume more power.

4-5-3   Volatile RAMS

Most present-day semiconductor RAMs are _volatile._ That is, all information is lost when the power is removed. This problem can be avoided by using a battery-maintained power supply during standby operation. Dynamic RAMs are often used for battery operation, since they draw less power than static RAMs. CMOS RAMs are particularly well suited for battery operation because of their low power consumption (as discussed in Sec. 3-1).

4-5-4   ROM circuits

The term "ROM" is applied to a wide range of devices in which fixed two-state information is stored for later use. Usually, the ROM is housed in a separate IC package containing a matrix of addressed locations. In a ROM, each bit in each address is *permanently set* to binary 1 or 0. ROMs are generally used to hold the program of microprocessor instructions and possibly other data constants, such as routines, tables, and so on. Unlike the RAM, the ROM is nonvolatile (the memory remains after power is removed), and the write function does not exist for a ROM (as discussed in Sec. 3-6-5).

Figure 4-17 shows the functional block diagram of a typical ROM. This units is described by the manufacturer as a 1024 × 8 ROM. This means

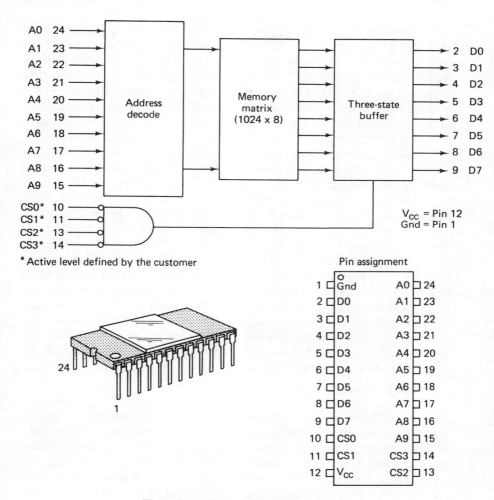

Figure 4-17 Motorola 1024 X 8 ROM.

that the matrix has 1024 addresses, each with 8 bits. Thus, 8-bit binary words or bytes can be read out of 1024 locations. The bytes appear on the data lines D0 through D7, connected to the data bus. Although the buffer is three-state, only the high impedance and output states are used. When chip select (CS) inputs are available, the permanent information stored at the selected address is passed to the data bus. When one or more of the chip select inputs are absent, the buffer remains in the high-impedance state and no data bytes pass from the matrix to the data bus. Thus, it is possible to shut off the ROM from the system by controlling only one of the lines or inputs.

Note that the user can define whether the chip select inputs are active high or active low. Also, note that the user must define the binary word to be stored at each address. Generally, the user defines the desired contents of the ROM by means of IBM cards or punched paper tape. The manufacturer then programs the ROMs (sets in the binary bit pattern at each address), usually by means of a mask for the final metalization step in IC manufacturer. This process is generally used where a large number of ROMs are required (typically 300 or more), but can be used for smaller numbers. As in the case of a RAM, the address to be read in or out of the ROM is selected by an address decoder, which, in turn, is controlled by the binary word on the address bus (lines A0 through A9).

There are two obvious problems for this method of custom ROM manufacture. First, the cost of the metalization mask (used for all the ROMs of a given program) is quite high. Thus, if only a few ROMs are needed, the cost per ROM is quite high. More important, the user rarely, if ever, knows the exact programming required for a ROM until after the program has been tested and debugged. That is, you do not know what binary word is to be located at which address until you have written the program, tested it, and found that the program works under all circumstances. There are several ways to overcome this problem, including the use of PROMs (programmable ROMs), EPROMs (erasable, ultraviolet PROMs), EAROMs (electrically alterable ROMs), and RMMs (read mostly memories), all of which are described in the following paragraphs.

4-5-5 PROM circuit

A PROM is shipped by the manufacturer with all bits at each address blank (or at binary 0, in most cases). The user then programs each bit at each address by means of an electrical current. Typically, when current is applied to a bit, a Nichrome wire is "fused" or opened by the current, making that bit assume the electrical characteristics of a binary 1. Bits to be at binary 0 are left untouched. The program in the PROM is then permanent and irreversible.

The programming can be done with simple switches and a voltage source, as shown in Fig. 4-18. In this circuit, the address switches are set to produce the binary word at the address terminals of the PROM (switches closed to provide 5 V for each binary 1, switches open for a 0). With the proper address selected, the rotary programming switch is set to each bit requiring a binary 1, in turn, and the fusing switch is pressed for about 1 second to fuse the Nichrome wire within the PROM. This process is very time-consuming and tedious. Also, if a mistake is made on even 1 bit in one address, the entire PROM is made useless for that program. There are automatic and semiautomatic PROM programming devices available to overcome this problem.

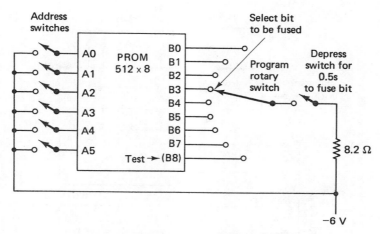

Figure 4-18 Basic PROM programming circuit.

4-5-6 EPROM circuit

Even though it is possible to program a large number of PROMs with automatic equipment rather quickly, there is still the problem of testing and debugging before a final program is obtained for the ROM. This problem is overcome by means of an EPROM in which information is stored as a charge in a MOSFET. Such EPROMs can be erased by flooding the IC with ultraviolet radiation. Once erased, new information can be programmed into the PROM in the normal manner. The erasure and reprogramming process can be repeated as many times as required. As shown in Fig. 4-19, an EPROM package is provided with a transparent lid that allows the memory content to be erased with ultraviolet (UV) light. For proper erasure, the semiconductor chip within the IC package must be exposed to strong UV light for a few minutes. Exposure to ordinary room light will take years to produce erasure; thus, there is no danger of the program being erased accidently. Even exposure to direct sunlight will not produce erasure for several days.

Except for the erasure feature, a typical EPROM is similar to a ROM in function, as shown by the block diagram of Fig. 4-19. In addition to the usual address inputs, data outputs, decoders, and buffers, and EPROM usually has a "program" input line. An appropriate signal on the program line permits information to be programmed into the selected address.

4-5-7 EAROM and RMM

These devices are a form of erasable PROM, but use electrical currents instead of UV light for erasure. The programming is similar to that of the

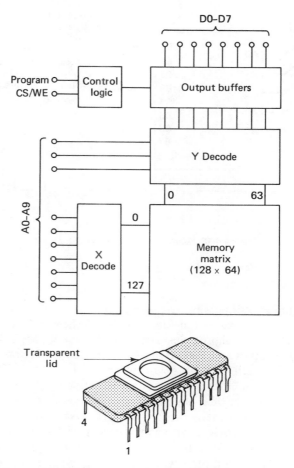

Figure 4-19   Motorola 1024 X 8 alterable (ultraviolet) ROM.

EPROM, but the erasure is a slow process and requires special circuits. For that reason, EPROMs are generally used for microprocessor-based systems rather than EAROMs and RMMs. However, this trend can change in the future.

4-6   INPUT/OUTPUT HARDWARE

The microprocessor and memories described in Secs. 4-4 and 4-5 can be connected to form a simple and almost complete microcomputer system. The missing element is an input/output device between the microcomputer hardware and the outside world, or peripherals. Without some I/O hardware, the transfer of data between the outside world and the microcomputer would be impossible.

Assume, for example, that a basic microcomputer (microprocessor, ROM, and RAM) is used with a conventional video terminal (keyboard and CRT video display), and that the data lines of the video terminal are connected to the microcomputer data buses. Information to be processed is typed on the keyboard and transmitted directly to the microcomputer data lines. After processing, the data bytes are returned directly to the video display. There are three basic problems with such an arrangment.

First, data bytes from the keyboard can easily appear simultaneously with data from the selected memory. Second, the video terminal and microcomputer would have no way of telling each other that they are ready to transmit or receive data. Third, there is no synchronization of timing between the microcomputer and the video terminal. That is, the microcomputer and video terminal clocks may be operating at different frequencies, phase relationships, and so on.

A further problem is created if the peripheral operates with serial data rather than the parallel data required for the microcomputer. Also, there is always the problem of *interfacing* electronic devices (different operating voltage levels, impedances, etc., as discussed in Sec. 3-6). These and other problems are overcome by means of an input/output IC. Most microprocessor manufacturers supply one or more I/O ICs for their systems. Typically, there is one I/O device for interfacing with parallel peripherals, and another for serial peripherals.

4-6-1 Typical parallel I/O IC

Figure 4-20 is the block diagram of a basic parallel I/O IC. The circuit is described as an 8-bit input/output port by the manufacturer, and consists essentially of an 8-bit register and 8-bit buffers, together with the control elements. The MODE control is used to program the device as an input port or output port. The MODE control is 0 for input and 1 for output.

Input port. When used as an input port, information is passed into the 8-bit register when pulses on the clock line are high. The clock also sets the service request (SR/SE) circuit and latches the data in the register. The SR output can be used to signal or flag the microprocessor that the data bytes from the peripheral are ready for processing. The CS1 and CS2 inputs are used to control the three-state buffers. The buffers are enabled when the CS1 and CS2 lines are high. This also resets the SR circuits (to flag the microprocessor that the data bytes have been passed).

Output port. When used as an output port, the buffers are enabled at all times, and information is passed into the 8-bit register (if CS1, CS2, and the clock are all high). The service request SR signal is generated when CS1 and

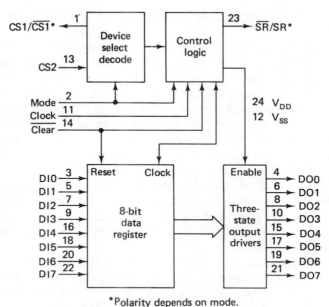

*Polarity depends on mode.

Figure 4-20   RCA COSMAC 8-bit I/O port.

CS2 are swinging low, and remains until the clock swings low. Stated another way, SR is generated when CS1 and CS2 go from 1 to 0, and remain until the clock line goes to 0.

Clear signal.   A CLEAR signal is provided for resetting the register and the service request circuit. The CLEAR function operates in both the input and output port modes.

From these descriptions, it can be seen that the circuit of Fig. 4-20 provides the basic I/O functions. Using the video terminal example, a data byte from the keyboard can be held in the register until the microprocessor is ready to accept new data. This condition is signaled or flagged to the microprocessor by the SR line. Then the buffers are enabled by the microprocessor and the data byte is passed to the microcomputer system. The buffers are set to the high-impedance state and the registers are reset. After processing in the microcomputer system, the data byte is placed in the registers, and this condition is flagged to the video terminal by the SR line. When the video terminal is ready to accept data for display on the CRT, the buffers are enabled by the video terminal, the byte is passed to the CRT display, the buffers are again returned to the high-impedance state, and the registers and SR line are reset.

4-6-2   Serial I/O IC

Operation of a serial I/O circuit is generally far more complex than that of the basic parallel I/O circuit just described. In addition to all the control, timing, and data-transfer functions, a serial I/O must also convert from parallel to serial, and vice versa for each exchange of information between the peripheral and microcomputer system. Because of this complexity, and because each manufacturer uses a somewhat different system for their I/O devices, we will not go into the details of serial interface here.

4-7   BASIC MICROCOMPUTER SYSTEM

Now that we have covered basic microcomputer hardware (microprocessor, ROM, RAM, and I/O ICs) and the circuits within these ICs, we are now ready to describe a simple, yet complete microcomputer system. Figure 4-21 is the block diagram of such an elementary microcomputer. One difficulty with understanding microcomputers is that there are many different types of microprocessors, and even more microcomputer system arrangements. As you will discover, there is no real standardization in hardware, software, or microcomputer system configurations.

Since it is difficult for anyone (even experienced technicians and engineers) not already familiar with computers of some type (standard, mini, or micro) to immediately grasp the operation of any microcomputer, we shall consider a generalized and simplified microcomputer that does not ex-

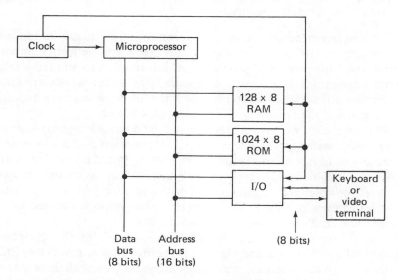

Figure 4-21 Block diagram of LENKMICROCOMP.

ist. Nevertheless, this microcomputer contains the basic principles upon which all microprocessor-based systems operate. Once you understand how this simplified microcomputer operates, you should have no difficulty in understanding any microprocessor-based system. Borrowing a technique used by microcomputer manufacturers, we shall call our system the LENK-MICROCOMP (LENK MICRO COMPuter).

4-7-1   Microcomputer system hardware

Our microcomputer consists of a microprocessor, a 128 × 8 RAM, a 1024 × 8 ROM, and an I/O device. The microprocessor is of special design but contains the usual registers, accumulators, counters, pointers, decoders, multiplexers, buffers, drivers, latches, ALU, and control circuitry. As is typical, the registers are used as temporary storage locations for data bytes or words, such as the accumulators in the ALU section, where the mathematical operations are performed. The counters are used to tally various items of information, such as commands (instructions) and locations (addresses) in memory where the information can be found. Decoders translate the instructions into electrical signals (pulses) for executing these commands and translate addresses into pulses. This permits the microprocessor to locate the required information and transfer it to the approximate destinations. Interconnecting all these elements are electronic and logic circuits (buffers, drivers, latches, flip-flops, etc.) that direct the signals along the appropriate paths.

4-7-2   Peripheral equipment

In most microcomputers there is a means of direct human-to-machine communication. The only exceptions to this are where the microcomputer system is used for control or processing of information to and from other electronic devices (such as when a microcomputer is used in industrial control). Human-to-machine communication is the job of peripheral equipment as is discussed in the remaining sections of this chapter.

For simplicity, we have chosen a manually operated keyboard device. This keyboard permits the operator to enter information into any location in the memory or to insert extra instructions. Our keyboard is in the form of a typewriter (similar to a Teletype or TTY instrument). With this arrangement, the typewriter can both insert and print out information. As an alternative, we can use a video terminal, where information is inserted by the keyboard and read out by the CRT video display.

The information fed to the microcomputer from the peripheral keyboard falls into two general classes. There are the *data bytes* (consisting of numbers, alphabetic letters, symbols, or combinations of all three that are to be processed), and the *instruction bytes* (the commands indicating how the data bytes are to be processed). Since the microcomputer's language is composed of binary numbers (in electrical pulse form), all information to be used must first be converted to numbers (or combinations of letters and numbers) and then the numbers must be converted to binary numbers.

In the LENKMICROCOMP keyboard, the conversion is done by means of a decoder. The keyboard decoder converts letters, numbers, and symbols from the keyboard into an ASCII code which appears at the keyboard output in the form of an 8-bit binary word or byte. This byte is in parallel form, and is applied to the microcomputer's 8-bit I/O IC. As an example, assume that the letter A on the keyboard is pressed. This is converted to a hex 41 by the keyboard decoder, and appears as binary 01000001 at the I/O terminals.

4-7-3 Microcomputer program

Microcomputers solve problems in a step-by-step manner. Since the microcomputer cannot think by itself (contrary to popular opinion), a *program* must be prepared that breaks the problem down into a series of sequential, logical, and simple steps. This is the task of the programmer. While an extended discussion on the art of programming (and it is an art, not a science) is not intended for this book, you must have some understanding (preferably a very detailed understanding) of the program if you are to learn how a microcomputer operates. For example, as discussed in Chapter 5, the first step in troubleshooting a microcomputer is to operate the system through its normal program and note any abnormalities in operation, sequence, or failure to perform a given step.

Since there are many variations among microprocessors, programs must be specifically developed for a particular microprocessor. The programmer must know the microprocessor language and the manner in which the microprocessor operates. In short, you must know the instruction set! The program that we shall consider here is for our hypothetical, general-purpose microcomputer, the LENKMICROCOMP.

4-7-4 The basic program

An as example of how a program may be prepared for the microcomputer, let us consider how far a freely falling body will fall in 8 seconds. The equation for this problem is $d = gt^2/2$, where d is the distance in feet, t the time in seconds, and g the acceleration due to gravity (32 feet per second for each second of fall). The programmer analyzes the problem and breaks it down into a sequential series of logical steps. The programmer then draws up a *flowchart,* which diagrams the sequence of steps to be taken in solving the problem. Such a flow chart is shown in Fig. 4-22.

Although the flowchart helps the programmer analyze the problem,

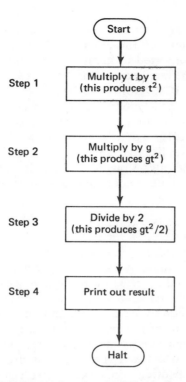

Figure 4-22 Basic flowchart for falling-body problem.

the microcomputer cannot use the flowchart as is. The microprocessor has its own language (machine language), which is based on binary numbers, not alphabetic letters or words. Accordingly, the flowchart must be converted to a program or sequential set of instructions (in machine language) which the microprocessor can follow. This conversion process is known as *assembly* and is discussed in Sec. 4-9. Some microprocessor manufacturers provide software programs, known as *assemblers,* that aid in the conversion process.

There are many ways in which manufacturers identify their instructions. First, some manufacturers call their instructions *operation codes* (or simply *op codes*). In some cases, each op code or instruction is identified by a binary number. This is the same binary number applied to the data inputs of the microprocessor to initiate an instruction or function. However, as discussed in Sec. 4-9, machine-language programming (using binary numbers) is very laborious and subject to error, except for very simple, short programs.

Most microprocessor manufacturers identify the instructions by an alphabetic abbreviation or *mnemonic,* by a numeric abbreviation (usually in hex), and by an assembly code or statement. Example of such identification for our LENKMICROCOMP are shown in Fig. 4-23. Note that the assembly code or statement is used for the convenience of the programmer when writing a program. However, only the numeric portion is used by the microprocessor (in binary form). Hex is used as a shorthand for the binary number.

Both the numbers representing the instructions and the numbers that constitute data are stored in the memory at particular addresses. Each instruction or data byte is stored at a separate address. The programmer must keep a record of the address of each instruction or data byte. Then, if the programmer wishes the microprocessor to obtain any specific instruction or data byte, the address is put on the address bus by the program, and the instruction or data byte stored at that address appears on the data bus.

Arrangement of instruction addresses.   There are several methods for arranging the instruction addresses in memory. Since the instructions follow in sequence, the address of the instructions may also be in sequential order. Thus, instruction 1 may be stored at address 0001, instruction 2 at address 0002, and so on. Before the start of operations, the address of instruction 1 (0001) is placed in a *program counter* register within the microprocessor. Then, as the microprocessor executes instruction 1, a signal is sent to the program counter, advancing the count by 1. The number in the program counter is now 0002, which is the address of instruction 2. The microprocessor obtains the address of the second instruction from the program counter (that is, the number in the program counter appears on the address bus). As the microprocessor executes instruction 2, the program counter advances to 0003, which is the address of instruction 3. This process continues as long as there are instructions to be carried out. The final instruction directs the microprocessor to stop.

Keep in mind that the instructions are *permanently programmed* into the ROM at the indicated addresses. Neither the instructions nor the order in

Alphabetic (Mnemonic) Representation	Numerical Representation		Meaning of Instruction
	Hex	Binary	
CAD	3A	0011 1010	Clear accumulator and add
ADD	3B	0011 1011	Add
SUB	3C	0011 1100	Subtract
MUL	3D	0011 1101	Multiply
DIV	3E	0011 1110	Divide
STO	3F	0011 1111	Store
PRT	7A	0111 1010	Print
HLT	00	0000 0000	Halt

Figure 4-23 Examples of typical op-codes for the LENKMICROCOMP.

which they appear can be altered (unless the ROM is erasable). An erasable or alterable ROM is used during development of the program. When the program is debugged and the desired order of instructions is determined, a ROM is permanently programmed to that order.

Arrangement of data addresses. There are several methods for arranging the data bytes in memory. One method is to store both the instruction and data bytes in each address. Then, when the microprocessor is directed to that address (that is, when the address word appears on the address bus), the microprocessor finds both the op code indicating an instruction and the data to be acted upon. Another method is to store each data byte at any address that is unoccupied and available. Using this system, each step of the program requires two machine cycles, one for instructions and one for data. Some microprocessors require three *machine cycles* for each program step. (As discussed in Sec. 4-8, a machine cycle usually consists of 8 consecutive bits of data.) Typically, the first machine cycle is for the op code or instruction, with the remaining two cycles for data. Keep in mind that data bytes are temporarily programmed into the RAM at the desired addresses. In our case, the data bytes are entered at the peripheral typewriter keyboard.

Note that with any of the systems for instruction and data addressing, the program does not contain the actual data but the address at which the data bytes are stored. For example, if the programmer wishes the microprocessor to add a certain number, the programmer indicates to the microprocessor for addition (the addition instruction word appears on the data bus) and gives the address of the data byte number to be added (the data byte address appears on the address bus). The microprocessor then obtains the data byte at the selected address, and adds the data byte number to whatever number appears in the microprocessor arithmetic register or accumulator.

Example of programming. As an example, let us program the falling-body program (using the flowchart of Fig. 4-22 and the instructions or op codes of Fig. 4-23). The data values involved are:

1. t, the time (in seconds) that the body is falling.
2. g, the acceleration (in feet per second of fall) due to gravity.
3. 2, the number by which gt^2 is divided.

As previously indicated, $t = 8$ and $g = 32$.

These data bytes may be stored at any unoccupied address in memory. For example, the data byte for t (8) can be stored at address 1003, data byte g (32) at address 1007, and data byte 2 at address 1008. The program listing will then appear as shown in Fig. 4-24.

This method of listing the program is often called an *assembly listing,*

Instruction operation data			Explanation
Address	Code	Address	
0001	3A	1003	CAD (clear and add). Erase any number remaining in the accumulator from a previous operation. Bring the contents of address 1003 (t) to the accumulator.
0002	3D	1003	MUL (multiply). Multiply the number in the accumulator by the number at address 1003 (the result is t^2).
0003	3D	1007	MUL (multiply). Multiply the number in the accumulator by the number at address 1007 (g) (the result is gt^2).
0004	3E	1008	DIV (divide). Divide the number in the accumulator by the number at address 1008 (2) (the result is $gt^2/2$.)
0005	3F	1138	STO (store). Store the number in the accumulator at address 1138.
0006	7A	1138	PRT (print). Print out data stored at address 1138.
0007	00		HLT. The microcomputer is directed to halt.

Figure 4-24 LENKMICROCOMP program for falling-body problem.

since it shows the program in assembly language (hex op codes, mnemonics, explanations, etc.) rather than in machine language. The differences in program listing methods are discussed further in Sec. 4-9.

Initially, the program counter is set to instruction address 0001 by a signal from the peripheral typewriter (via the keyboard decoder, I/O IC, and the data bus and/or control lines). The instruction at that address (which happens to be op code 3A) tells the microprocessor to clear the accumulator to zero and then add the number found at data address 1003 (which is decimal 8, or t). In some microprocessors, this operation can be done in one machine cycle, since part of the byte at address 0001 is the op code (3A) and part is the address (1003) at which the data byte to be addressed resides. This second part of the byte is described as the operand in some microprocessor literature (since the second part is to be operated on by the instruction or op code).

In our microcomputer, the operation is done in two machine cycles (one for the op code, one for the data or operand address). However, the program counter in our microprocessor advances only one step (to instruction address 0002), even though two machine cycles are required. The instruction at address 0002 (which is op code 3D) tells the microcomputer to multiply the number in the accumulator by the number found at data address

1003. This is the same data address used in the previous step and contains a data byte equal to decimal 8. The result in the accumulator is now 8 × 8, or 64, or t^2. Again, two machine cycles are used (one for instruction and one for data), and the program counter advances to instruction address 0003.

The instruction at address 0003 (again op code 3D) tells the microprocessor to multiply the number in the accumulator by the number found at data address 1007 (which is decimal 32, or g). The result in the accumulator is now 64 × 32, or decimal 2048, or gt^2, and the program counter advances to instruction address 0004. The instruction at address 0004 (now op code 3E) tells the microcomputer to divide the number in the accumulator by the number found at data address 1008 (which is decimal 2). The result is now $gt^2/2$, or decimal 1024, which is the answer to our problem (the body will fall 1024 feet in 8 seconds). The program counter advances to instruction address 0005.

The instruction at address 0005 (now op code 3F) tells the microprocessor to store the number appearing in the accumulator at address 1138. The program counter advances to instruction address 0006, which tells the microprocessor to print out the result at address 1038 on the peripheral typewriter (via the data bus, I/O IC, and keyboard decoder). The operator thus has a permanent record of the answer. In the case of a video terminal, the answer appears on the CRT display. After the printout, the program counter advances to instruction address 0007, which directs the microprocessor to halt.

4-7-5   Special program instructions

In addition to the routine instructions for our LENKMICROCOMP system, typical examples of which are shown in Fig. 4-23, there are a number of instructions that help make the microcomputer a flexible, decision-making machine. These are the *branch* instructions (also known as jump, skip, or transfer instructions) shown in Fig. 4-25. These instructions direct the microcomputer to leave the main program at some designated point and, under proper conditions, branch or jump to some other designated point in the program. This is sometimes known as a branch or subroutine *call*.

For example, assume that op code 20 and address 0033 appear at a cer-

Alphabetic (Mnemonic) Representation	Numerical Representation Hex	Binary	Meaning of Instruction
BRA	20	0010 0000	Branch, unconditionally
BRN	21	0010 0001	Branch, on negative
BRP	22	0010 0010	Branch, on positive

Figure 4-25   Examples of branch op codes for the LENKMICROCOMP.

tain point of the program, say at instruction address 0011. The instruction (op code 20) tells the microprocessor to leave the main program at this point and branch or jump to address 0033. However, the program counter first increments by 1 to the address 0012. The contents of the program counter (0012) are then put into an address in an unoccupied area of memory, usually called the *stack*. The stack thus saves the address of the instruction to be executed after the branch routine is completed.

With the address of the next step in the main program safely stored in memory, the program counter goes to the address specified by the branch instruction (0033 in our case). After the instruction at 0033 is performed, the microprocessor follows, in sequence, instructions at addresses 0034, 0035, and so on, until the last address of the branch instruction is reached. This last address usually contains an instruction to return to the main program. Such an instruction need specify no address. When the microprocessor receives a return instruction, the microprocessor replaces the current contents of the program counter with the address stored in the stack (0012). This causes the microprocessor to resume execution of the original program at the point immediately following the original branch instruction.

Conditional and unconditional branches. We have just described an *unconditional* branch instruction. The other two branch instructions shown in Fig. 4-25, BRN (op code 21) and BRP (op code 22), are examples of *conditional* branch instructions. If, for example, op code 21 and address 0033 appear in the program, it means that if the number in a certain counter or register is negative, the microprocessor must branch to address 0033, perform the instruction, and proceed (in sequence) from there. If the number is not negative, the microprocessor is to ignore the branch instruction and proceed with the original program. The BRP instruction is the same as the BRN except for branching it to occur if the number is positive. Otherwise, the branch instruction is ignored.

4-7-6 Microcomputer operation cycle

Once the program has been entered (or *loaded*) into memory (that is, with instruction permanently programmed into the ROM, and data bytes temporarily entered into the RAM via the peripheral typewriter), the microcomputer is ready to go automatically through the program cycle. Each cycle has two alternate phases; the *instruction* phase (or instruction *fetch* as it may be called) and the *execution* phase. After these two phases have been completed, the cycle is repeated. The duration of each phase is determined by a fixed number of clock pulses. Typically, there are eight clock pulses per machine cycle. Each phase may require one or more machine cycles.

The combination fetch and execution of a signal instruction is referred to as an *instruction cycle.* The portion of a cycle identified with a clearly defined activity is called a *state.* The interval between clock pulses is referred to as a *clock period,* as discussed in Sec. 4-8. Typically, one or more clock periods are necessary to complete a state, and there are several states in a cycle. The flow diagram of the LENKMICROCOMP operation cycle is shown in Fig. 4-26.

Instruction or fetch phase.   The first instruction fetch is initiated by a start command. In our case, this command is initiated by the peripheral typewriter. The start command sets the first instruction address (0001) into the program counter. The instruction phase of the operation cycle starts when this address is transferred to the address register.

     Note that as information is transferred from one storage location to another, the information is not erased from the original location unless specifically so ordered. Thus, in our example, the address of the first instruction word appears in both the program counter and the address register. On

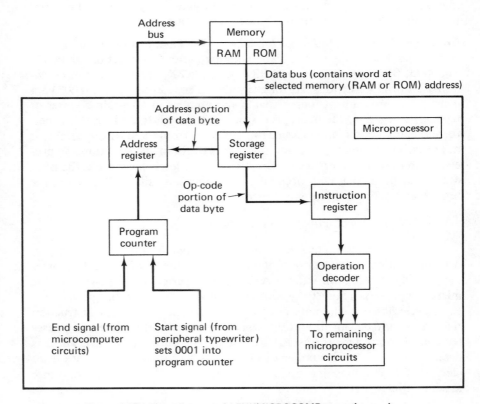

Figure 4-26   Flow diagram of LENKMICROCOMP operation cycle.

the other hand, when new information is placed in a storage location, any previous information stored at that location is first erased.

The output of the address register is applied to memory via the address bus, as shown in Fig. 4-26. The word at the selected address in memory is transferred to the storage register via the data bus. From the storage register, the op-code portion of the word is transferred to the instruction register. Here, the word is decoded by the operation decoder, and the microprocessor circuits (ALU, etc.) perform the indicated instruction. Also from the storage register, the address portion of the word is transferred to the address register, first erasing the information previously stored there. This completes the instruction phase of the cycle.

Execution phase.   The execution phase starts when the data address placed in the address register during the instruction phase is applied to memory via the address bus. The data word at the selected address in memory is transferred to the storage register via the data bus. The data byte is then processed in accordance with the instruction (still in the instruction register). When the processing is finished, an *end signal* or *processing complete* signal advances the program counter by 1 (to 0002 in our case). The program counter now contains the address of the next instruction, and the instruction fetch or phase can start.

This completes the execution phase of the operation cycle. The cycle is repeated over and over again, the instruction phase alternating with the execution phase for each cycle, until the entire program is complete. In the case of a branch operation, the op-code portion of the instruction word is that of the branch instruction, and the operand portion is the address of the next instruction to be followed if branching is to take place. Under such conditions, this address is placed in the program counter, replacing the address already there, as described in Sec. 4-7-5. The microcomputer then follows the branch program until a return to the original program is instructed.

We have just described the complete operation cycle of the LENKMICROCOMP. Different microcomputers may have different methods for going through the full cycle, but the basic principles of operation are essentially the same for all microcomputers.

4-7-7 Subroutines, libraries, and nesting

Practically every instruction to the microprocessor involves a routine series of steps. For example, to add two 8-bit numbers, the LSD must be added first. If there is a carry, the carry must be added to the next-significant bits. Then these bits must be added. If there is another carry, this carry must also be added to the MSD, and then they must be added. Although this series of calculations is complete within itself, the series may only be part of a larger program.

Since the microprocessor can only follow instructions, this routine series of steps must be included in the program. However, to save the user's time, this series of steps is listed in the microprocessor literature. Such listings or general-purpose set of instructions are called *libraries, routines,* or *subroutines.* Often, the term "libraries" is given to a group of routines.

A typical subroutine might include all the steps necessary to find square root, sines, cosines, logarithms, and so on. If these types of calculations are repeated throughout the program, the programmer may prepare a special program for each of them and store the program in memory. Usually, subroutines are handled as a branch or jump. Then, when the programmer calls for a routine, the microprocessor is given the branch op code and the address of the first branch instruction, as discussed in Sec. 4-7-5.

Subroutines are often *nested.* That is, one subroutine calls a second subroutine. The second may call a third, and so on. This is acceptable as long as the microcomputer has enough capacity to store the necessary return addresses, and the microprocessor is capable of doing so. The maximum depth of nesting is determined by the depth of the stack (or reserve memory). If the stack has a space for storing three return addresses, three levels of subroutines may be accommodated.

4-7-8 Interrupts

Most microprocessors have some provision for interrupts. Generally, interrupts are used for communications with peripheral equipment. For example, assume that the LENKMICROCOMP is processing a large volume of data, portions of which are to be printed on the peripheral typewriter. The microcomputer can produce one 8-bit data byte output (or one alphabet character) for each machine cycle, but it may take the typewriter the equivalent of several machine cycles (or several dozen) to actually print out the character specified by the data byte. The microprocessor could then remain idle, waiting until the typewriter can accept the next data byte.

If the microprocessor is capable of interrupts, the microprocessor can output a data byte, then return to data processing. When the typewriter is again ready to accept the next data byte, the typewriter can request an interrupt. When the microprocessor acknowledges the interrupt, the main program is suspended, and the microprocessor automatically branches to a routine that will output the next data byte. (This is known as *servicing* the peripheral device, or servicing the interrupt.) After the data byte is delivered to the typewriter and the corresponding character is printed, the microprocessor continues with main program execution. Note that this is, in effect, quite similar to a branch or subroutine call (discussed in Sec. 4-7-5), except that the jump is initiated externally (by the typewriter) rather than by the program.

There are more complex interrupt schemes, particularly where several peripherals must share the same microprocessor. Usually, such interrupts involve assigning priority levels (prioritizing) the peripherals. One approach (called the *polled method*) is for the microprocessor to check each peripheral, in turn, on a given priority basis, for a service request. The polled method is generally wasteful, since part of the main program is used up whether the peripherals need service or not. It is generally more effective for the microprocessor to stop only when the peripheral sends an interrupt. The peripheral can still be assigned priorities. For example, if peripheral A is first priority and B is second priority, A will be serviced before B if both peripherals make a service request simultaneously. B will be serviced first only if A is not making a request.

4-8 MICROCOMPUTER TIMING AND SYNCHRONIZATION

It is obvious that all the system functions described in Sec. 4-7 and shown in Fig. 4-26 must be synchronized as to time. For example, if a data byte is to be entered into memory at a particular address, that (and only that) data byte must be on the data bus when the desired address byte is on the address bus. If the two bytes (data and address) are not synchronized exactly, the data byte will be entered at the wrong address (or at no address). This problem is overcome by timing within the microprocessor. Registers are opened and closed at exact time intervals (by the clock pulses) to accomplish the desired results.

System timing is primarily an internal function of the microprocessor. As in the case of microprocessor instructions, you cannot change the time relationships or operating cycles of the microprocessor. However, you can change system speed or operating frequency (which is determined by the clock pulse frequency). Clock frequency, internal or external, is generally controlled by a quartz crystal, identical to those used to control frequency in communications transmitters. If a 1-MHz crystal is used, the clock pulses will be approximately 1 μs; with a 3-MHz crystal, the clock pulses will be about one-third of this, or about 333 ns. However, if it takes eight clock pulses to complete a certain function or instruction at one speed, it will take eight clock pulses to complete the same function at any other speed.

The time relationships and synchronization are shown by *timing diagrams* such as Fig. 4-27. The clock pulses are shown at the top and bottom of the diagram. There are two sets of eight clock pulses. Thus, each clock pulse represents 1 bit in an 8-bit byte. Some functions are accomplished within the first byte; other functions require two bytes. Note that there is no reference to operating speed on the clock pulse, but that one clock cycle is shown as a time interval IT (at clock pulse 3 on the top line). This is known as

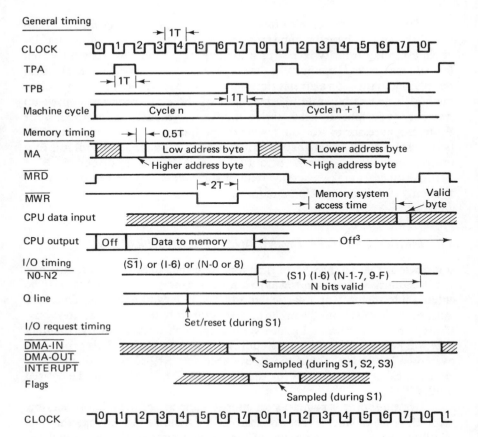

Figure 4-27 Timing diagram for basic RCA COSMAC microcomputer system.

a *clock period.* If the clock frequency (set by the crystal or other external clock pulse source) is 1 MHz, the time interval or clock period IT is approximately 1 μs.

From a programming standpoint, the timing intervals and relationships are not that critical. It is generally more important that the programmer know that it takes one or two bytes to accomplish a given function or instruction. However, from a design or troubleshooting standpoint, timing synchronization is critical. For example, when considering a design, the maximum operating speed of the microprocessor and the number of bits (or bytes) required for each of the instructions determines the number of functions that can be done in a given time.

From a troubleshooting standpoint, the timing can be even more critical. For example, as discussed in Chapter 5, one of the standard troubleshooting techniques for microprocessor-based systems is to display the microprocessor signal pulses (as many as possible) on an oscilloscope or logic analyzer. In effect, the timing diagram of Fig. 4-27 (or a significant portion of it) is displayed on the oscilloscope or analyzer. Then the time relationships are compared. As an example, note that the timing pulses TPA and TPB (sent from the microprocessor to control peripherals and I/O circuits) occur at 8-bit (1-byte) intervals. However, TPB occurs 5½ bits (or 5½ cycles) after TPA. If either TPA or TPB is absent or abnormal (wrong time relationship—say 3 bits apart instead of 5½), this would pinpoint a fault in the system.

It is essential that you understand the time relationships of a microprocessor, just as you must understand the instruction set, to make full use of a microprocessor (or to troubleshoot a microprocessor-based system). Such timing diagrams are found in the microprocessor literature. It is not uncommon to have several timing diagrams, one for each major function or group of functions. We will not go into a full discussion of timing here. However, one point of interest should be noted in Fig. 4-27. Each machine cycle consists of eight clock pulses (or a full 8-bit byte), and each instruction or function requires two or three bytes (or two or three machine cycles, whichever term is used by the microprocessor manufacturer). This relationship among clock bits, machine cycles, and instruction or function timing is common to many microprocessors.

4-9 *MACHINE LANGUAGE VERSUS ASSEMBLY LANGUAGE*

A microprocessor may be programmed by writing a sequence of instructions in binary code which the microprocessor can interpret directly. For example, referring back to the instructions of the LENKMICROCOMP shown in Fig. 4-23, the instruction for addition is binary 00111011. If this byte is programmed into memory and then appears on the data bus when the microprocessor data port is open, the microprocessor will add (probably the contents of one register to another). This is *machine language programming* and is useful only where the program to be written is small.

At best, writing a program in machine language is a tedious task, subject to many errors. The task of writing a program can be speeded up and error minimized when hex is used instead of binary (both hex and binary are shown in Fig. 4-23). However, hex coding has two major limitations. First, hex coding is not self-documenting. That is, the code itself does not give any indication in human terms of the operation to be performed. The user must learn each code, or constantly use a *program reference card* or *sheet* to convert. Second, and more important, hex coding is *absolute*. That is, the program will work only when stored at a specific location in memory. This is because branch or jump instructions in the program make reference to specific addresses elsewhere in the program.

Consider the example shown in Fig. 4-28, which is a listing of an eight-step program written in machine code (but in hex format). Steps 0007 and 0008 make reference to step (or instruction address) 0004. If the program is moved, step 0008 must be changed to refer to the new address of step 0004.

4-9-1 Assembly language

The problems just discussed can be overcome by writing the program in *assembly language,* where alphanumeric symbols are used to represent machine language codes, branch addresses, and so on. For example, the instruction to increment the contents of register 0 becomes INC R0 instead of hex 18, giving the user the meaning of the instruction at a glance. Our example program of Fig. 4-28 can then be written in assembly language as shown in Fig. 4-29.

The use of assembly language makes it much easier to write a program but it results in a program that is useless to the microprocessor (which operates only with binary numbers). One way to overcome this problem is to write the program in assembly language but to include the hex codes (as

Step Number	Machine Code	Explanation
0	1011 1000	Load decimal 32 in
1	0010 0000	register R0
2	1011 1010	Load decimal 5 in
3	0000 0101	register R2
4	0000 1001	Load port 1 to accumulator
5	1111 0000	Transfer contents of accumulator to register addressed by register 0
6	0001 1000	Increment R0 by 1
7	1110 1010	Decrement register 2
8	0000 0100	by 1; if result is zero, continue to step 9; if not, go to step 4
9	—	
10	—	

Figure 4-28 Basic eight-step program written in machine code arranged in hex format.

Step Number	Hex Code		Assembly Code
0	B8		MOV RO, #32
1	20		
2	BA		MOV R2, #05
3	05		
4	09	INP:	IN A, P1
5	F0		MOV @RO, A
6	18		INC RO
7	EA		DJNZ R2, INP
8	04		

Figure 4-29 Basic eight-step program written in assembly language, but including hex code.

shown in Fig. 4-29). Then, when you enter the program into the microcomputer, it is relatively easy to convert from hex to binary. Of course, this requires that you remember both the hex code and the assembly code for each instruction of the microprocessor. Although it is not a bad idea for you to learn all of the hex codes, it does require more remembering (or looking up) on your part, and thus increases the chance for error.

A more convenient method is to assemble programs using an *assembler,* which is a program that converts from one code to another. In our case, the assembler converts from a verbal-type assembly code to binary numbers. In practice, the assembler program is entered into another computer, you write the program for your microcomputer on paper (known as the *source program*), and you enter your microcomputer program into the other computer (known as the *host computer*), which prints out a program (usually on paper tape or disks or possibly on cassette tapes) in binary form. This program is known as the *object program,* and can be entered directly into your microcomputer from the paper tape, disk, or cassette.

When a different computer is used for assembly, the program is known as a *cross-assembler.* It is also possible to use *time-sharing* services for assembly, and there are special devices and equipment developed by manufacturers for assembly of their microcomputer programs.

4-10 PERIPHERAL EQUIPMENT FOR DIGITAL COMPUTERS

In a microcomputer, the primary function of peripheral equipment is to translate human instructions and data into microprocessor language (binary data bytes) and from microprocessor language into a form suitable for readout. In effect, the peripheral equipment (in conjunction with the microcomputer I/O IC) reconciles the "outer world" with the microcomputer. There are two major reasons for this reconcilation or translation. First, the outer world rarely expresses anything in binary data bytes.

(Generally, the outer world uses numbers, letters, words, etc.) However, the microprocessor always uses binary data bytes. Second, the microprocessor operates at extremely high speeds when compared to the outer world. For example, in the fraction of a second that it takes to strike a key of an input typewriter, a microprocessor can perform hundreds or thousands of operations.

4-10-1 Peripherals in data processing

To take advantage of available microprocessor speed when used in data-processing applications, the information to be processed is prepared in advance and stored. Common storage forms include paper tape, punch cards, and magnetic tape, drums, or disks. The basic relationship between a data-processing microcomputer and peripheral devices is shown in Fig. 4-30.

Note that the terms *on-line* and *off-line* are used to indicate *direct* and *indirect* connection to the microcomputer. For example, a paper-tape punch that is operated by an electric typewriter is an off-line device for indirect input. The paper-tape reader that converts holes in the paper tape into pulses

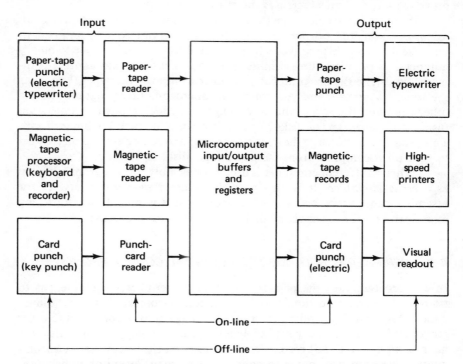

Figure 4-30 Basic relationship between the microcomputer and input/output accessories in data-processing applications.

applied to the microcomputer is an on-line device for direct input. Information is put into the off-line devices by human operators. The on-line devices then automatically convert the information into data bytes (in pulse form) for application to the microcomputer. On-line devices require only that the human operator load the tapes or cards and turn on the unit.

With a data-processing system such as the one shown in Fig. 4-30, the microcomputer does not have to wait for the operator to type out messages, since the microcomputer can read directly from the tapes or cards. Large quantities of tapes or cards can be made in advance by many operators, handling different problems on separate typewriters. The microcomputer can look at all the tapes or cards from the appropriate reader. After processing, the microcomputer directs the on-line tape or card punch (or magnetic recorder) to store the processed data. The tapes or cards containing the processed data are then read out on output printers or typewriters (or other visual display devices).

These output readout devices are not as fast as the microprocessor. However, the devices are very fast in comparison to human operators and have a chance of keeping pace, since many microprocessor calculations require repeated steps.

4-10-2 Peripherals in real-time processing

Not all microcomputers are used for data processing. In fact, most high-volume data-processing systems still use larger computers and minicomputers. Many microcomputers operate on a real-time basis, where the microcomputer stands ready to solve problems presented by the human operator. Where human operators must command a microcomputer, a manual keyboard (as found on an electric typewriter or TTY) is the simplest and most common form of input/output device. The commands are typed and applied directly to the microcomputer (in the form of data bytes). The processed output signals from the microcomputer then "type out" the answers on the same keyboard.

4-10-3 Peripherals in process control applications

Many microprocessor-based systems receive continuously changing inputs and produce corresponding readouts. Microprocessor-based systems used to monitor (and control) industrial processes are an example. There are generally no off-line devices with such systems. Instead, these microprocessor systems receive inputs from *transducers,* which convert such factors as process temperature, flow rate, and volume into electrical voltages. The voltages, which are *analogs* of the temperature, flow, and volume, are converted into data bytes by an analog-to-digital circuit (Sec. 3-3). The automatic readout from the microprocessor system, in addition to supplying a permanent record on a typewriter or printer, is also used to control the industrial process. At the microcomputer output, the data bytes are converted to a corresponding electrical voltage by a digital-to-analog circuit (Sec. 3-3). The output electrical voltages are then applied to control voltages, switches, motors, valves, and so on, as needed to control the industrial process. For a full analysis of industrial control systems, your attention is invited to the author's best selling *Handbook of Controls and Instrumentation,* Prentice-Hall, Inc., 1980 (Englewood Cliffs, N.J.).

4-11   ELECTRIC TYPEWRITER OR TTY TERMINALS

Electrical typewriters and TTYs provide both input and output for a microcomputer. Generally, the same I/O IC is used for both input and output functions.

4-11-1   Basic input circuit

Figure 4-31 shows the input relationship between a peripheral typewriter or TTY and a typical microcomputer. With any peripheral typewriter system, there must be circuits between the keyboard and the microcomputer input that (1) convert the data instructions into bytes compatible with the microcomputer circuits, and (2) store the bytes until they can be entered into the microcomputer circuits without disrupting normal operation. This procedure is sometimes referred to as a *service routine.* The conversion is done by a decoder circuit, while the storage and timing is done by registers and buffers. The present design trend is to include the register and buffer function in the I/O IC of the microcomputer, with the decoder function in the typewriter.

Note that the data bytes are transmitted in both parallel and serial form between the peripheral and the microcomputer. The parallel method is the fastest and requires the simplest I/O circuit in the microcomputer. However, TTY terminals generally transmit their data bytes over a single telephone line (or pair of lines). Eight lines, plus any control, flag, service request, or interrupt lines, are required for an 8-bit parallel system. Thus, TTYs (and all serial data transmission devices) require a special serial I/O IC (often called an *asynchronous* or *universal* I/O). Most major microprocessor manufacturers produce at least one serial I/O, as discussed in the remaining sections of this chapter.

TTY serial data transmission.   While on the subject of TTY operation, it is conventional to transmit serial data in a fixed format similar to that shown in Fig. 4-32. When a user strikes a key on the TTY keyboard (an "M", for example), the information denoting that character is converted to its ASCII

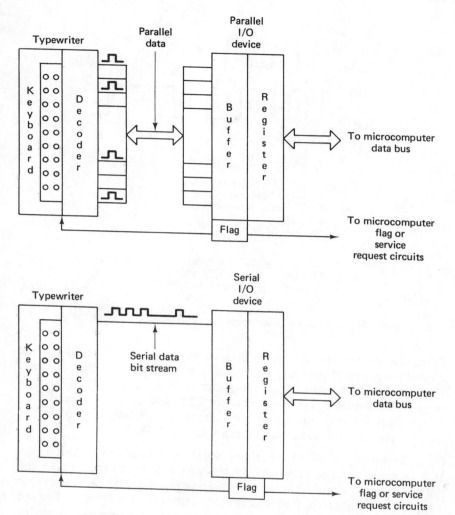

Figure 4-31 Input relationship between a peripheral typewriter and a microcomputer for both serial and parallel operation.

code (4D in hex) and appears at the TTY output as a *serial data bit stream*. Note that each of the character data bits (b0 through b6) is identified on Fig. 4-32. The character (the letter M in this case) is framed by a start bit B and 2 stop bits FF. By convention (or *protocol*), 2 stop bits are used for data transmitted at 10 characters per second, and 1 stop bit for higher data-transmission rates.

A parity bit P is also shown in Fig. 4-32. The parity bit is a 1 only if the

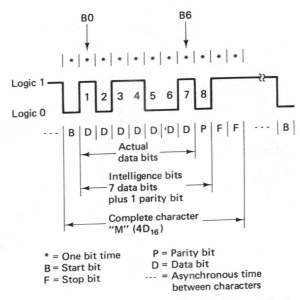

* = One bit time   P = Parity bit
B = Start bit   D = Data bit
F = Stop bit   --- = Asynchronous time
                           between characters

Figure 4-32   TTY serial data transmission format (showing character "M").

7 data bits contain an odd number of 1s. Hence, the total number of 1s in the 8 intelligence bits (7 data bits plus 1 parity bit) is always an even number. This convention is called the even-parity option for ASCII-coded data transmission (which is, in general, used for TTYs and most of the serial data-transmission devices). Since there are 4 bits (an even number) in the ASCII code for the letter M, the parity bit is a 0 in Fig. 4-32.

Input timing sequence.   With either serial or parallel transmission, the data/instructions must pass from the typewriter to the microcomputer circuit in a precisely timed sequence. As shown in Fig. 4-31, the decoder output is applied to a register/buffer combination. The buffer is opened and closed by pulses or control signals from the microcomputer, in response to *service requests* (also called flags, interrupts, etc.) from the typewriter. This reconciles the high speed of the microcomputer with the slow speed of the typewriter. For example, assume that the buffer can be opened and closed 10,000 times per second and that it takes 1s for the typewriter key to be pressed and released. No matter what time the data byte from the keyboard starts and stops, the buffer/register has thousands of times to read in, store, and read out the data byte to the microcomputer.

This combination of service request signals and timing signals is necessary to prevent undesired conflict between the typewriter and microcomputer circuits. Assume, for example, that the microcomputer is performing a series of mathematical calculations on data bytes previously entered by the keyboard and that the operator starts to enter more data. In the normal sequence, a service request is sent by the typewriter. Generally, a service request is sent for each data byte, or each time a typewriter key is struck. The service request indicates to the microprocessor that there is new information to be entered into the buffer. If old information in the buffer has been read out to the microcomputer, the buffer is reset to zero, and the new data bytes are entered and read out in the microcomputer.

Now assume that the microcomputer is not ready to accept new information (previous data bytes are still being processed). Then the buffer accepts the new data bytes, but does not pass them to the microcomputer until the circuits are ready. Keep in mind that since the microcomputer operates at such a high speed, the operator does not have to "wait" until the computer is finished before striking the next key.

4-11-2 Basic output circuit

Microcomputer output circuits are essentially the reverse of input circuits. The output circuits to an electric typewriter or TTY must (1) convert the processed data bytes to a form (usually numbers, letters, words, etc.) that is compatible with the typewriter keys, and (2) synchronize the high-speed microprocessor output with that of the lower-speed typewriter. As in the case of input, the conversion is done by a decoder circuit (usually within the typewriter), whereas the synchronization is done by the I/O buffer/register, which operates in response to a combination of microcomputer and typewriter signals.

Figure 4-33 shows the output relationship between a peripheral typewriter or TTY and a typical microcomputer. In the readout mode, the keys are operated by electrical voltages from the decoder. When the buffer/register is opened, the data byte is applied to the decoder, which, in turn, applies the operating voltage to the appropriate typewriter key. This causes the appropriate key to strike and print out the corresponding character (number, letter, symbol, etc.). In the normal sequence, a service request is sent by the typewriter. This indicates to the microcomputer that the keyboard is ready to receive data from the buffer. The register is set to zero, the new data byte is entered from the microcomputer, the buffer is opened, and the data byte is passed to the decoder. In turn, the decoder operates the corresponding key.

It is obvious that the microcomputer can solve problems faster than the keyboard can type them out. This means that two conditions must occur. First, the output buffer must be held in the existing state long enough for the key to be struck. Second, the normal routine of the computer must be interrupted so that the answers to several problems are not computed while the

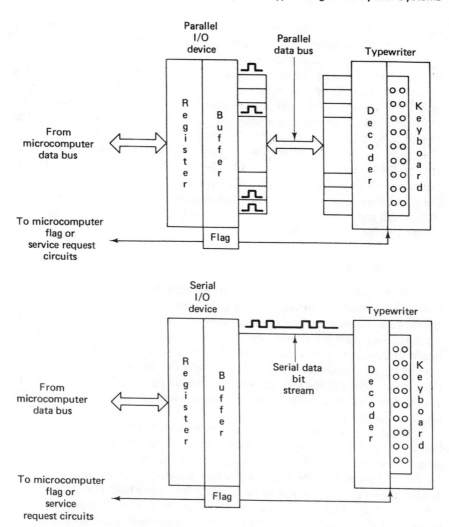

Figure 4-33 Output relationship between a peripheral typewriter and a microcomputer for both serial and parallel operation.

output buffer is held open. In actual practice, the interruption occurs at some convenient point in the microcomputer operation (usually at the end of a timing cycle).

4-11-3 Typical keyboard circuits

Figure 4-34 shows a portion of some typical keyboard circuits. The portion shown is from the keys through an encoder IC402 (rather than a decoder) and buffers (IC403) to a parallel-to-serial converter (called a UART, or universal asynchronous receiver transmitter). As shown, the keyboard contains a 52-key ASCII set of keys.

The ASCII keyboard is a matrix of single-pole, normally open switches that are scanned by IC402. The X2 through X9 outputs from the keys are scanned by a series of pulses. When a key is pressed, one of these pulses appears at one of the eight Y inputs. After an appropriate period to allow for the key to stop bouncing (known as the debounce period), the encoder IC402 recognizes the pulse as a valid key depression, decodes the X and Y pulse data into the appropriate 7-bit ASCII word, latches the word in buffer IC403, and puts a logic 1 on the data strobe output (pin 13). This data strobe output indicates to the UART that there is a valid data byte present in the buffer.

The negative-true logic output of encoder IC402 is inverted by inverter/buffer IC403 to provide positive-true logic to the UART. When the output enable line (OE) is brought to a logic 0, the three-state buffers are turned on and the ASCII data bits are applied to the UART. A logic 0 on the OE line also resets the data strobe output of IC402 to indicate that the data byte in the buffer has been read. The serial output from the UART is then passed out of the typewriter to the microcomputer through serial input/output circuits as described in Sec. 4-11-4.

Note that the encoder IC402 is operated by clock pulses from a *baud rate generator*. The term "baud" as used in serial data transmission or communications is a measure of the transmission rate. Loosely, baud indicates *bits per second,* but usually includes the characteristic framing start and stop bits shown in Fig. 4-32. The higher the baud rate, the faster the serial data transmission rate. Typical baud rates are 50, 75, 110, 134.5, 150, 200, 300, 600, 1200, 1800, 2400, 4800, and 9600. The baud rate for the keyboard of Fig. 4-34 is 4800.

4-11-4   Serial input/output

The serial input and output of many peripheral typewriters (and other terminals) can be jumper-wired for one or more standard data transmission configurations. The three most commonly used serial I/O configurations found on peripherals of all types are TTL (Sec. 3-1-4), EIA (Electronic Industries Association), and TTY (also known as *20-mA current loop).* The two logic states of a serial input or output are called a *mark* and a *space.* For TTL serial I/O, a mark is defined as a TTL logic 1. For EIA, a mark is a minus voltage. For TTY, a mark is 20 mA flowing in the current loop between the peripheral and the I/O. For TTL, a space is defined as a TTL logic 0. For EIA, a space is a positive voltage. For TTY, a space is zero current in the 20-mA current loop. The idle state of both the input and output is defined as a mark.

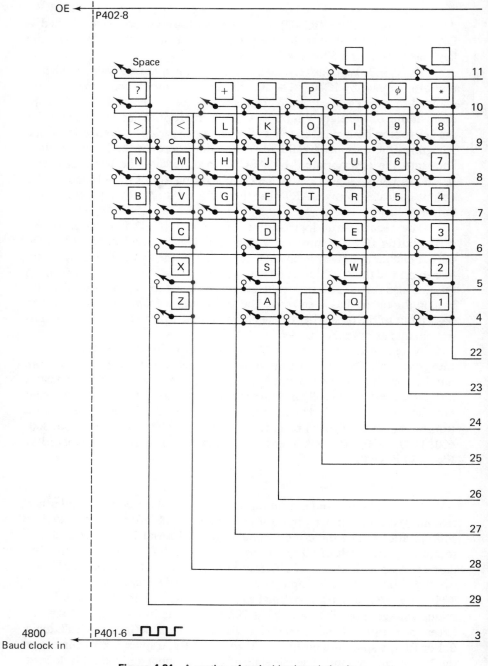

Figure 4-34 A portion of typical keyboard circuits.

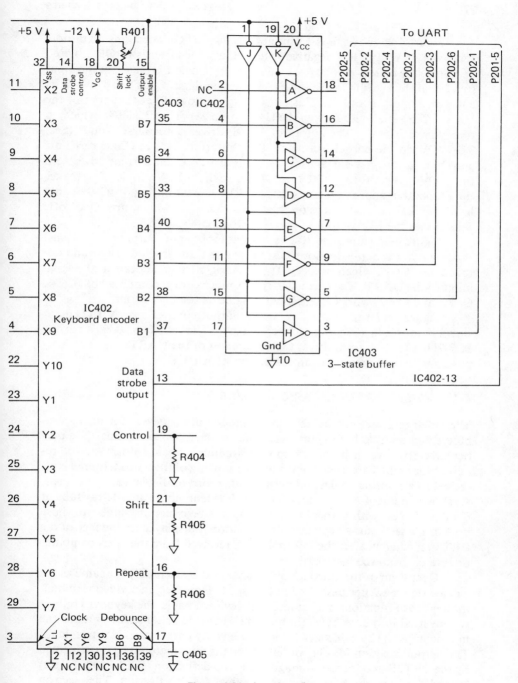

Figure 4-34 (*continued*).

Figure 4-35 shows the EIA input/output circuit connections. In the idle state, the EIA input is a negative voltage that keeps transistor Q603 biased off. Since the collector voltage of Q603 is 5 V, a logic 1 is placed at the input (pin 20) of the UART. When the EIA input goes high, Q603 turns on and its collector voltage goes low. This places a space (logic 0) at the input of the UART. The serial output from the UART (pin 25) drives IC609A, which, in turn, drives IC609B. The output of Q609B drives optocoupler IC604 (Sec. 3-5-7). When the output is a mark, the phototransistor in IC604 turns off and Q602 (a constant-current source) turns Q601 on. The current flows from the + 12-V supply through R631 and Q601 to the − 12-V supply. This makes the EIA output approximately − 10 to − 12 V. As the output of IC609B goes low, the LED in IC604 turns on the phototransistor, which turns Q601 off. This causes the EIA output to go to approximately + 12 V.

Figure 4-36 shows the 20-mA current loop input/output connections. When a mark is applied, a loop current passes through diode D601 and optocoupler IC605, which turns Q603 off. A logic 1 then appears at the serial input to the UART. The current-loop output circuitry, composed of IC604, Q601, and Q602, works essentially the same as for EIA ouput.

Figure 4-37 shows the TTL input/output connections. A TTL input signal is coupled to pins 9 and 10 of IC609C. The output of IC609C drives IC609D, which, in turn, drives the serial input of the UART. The TTL output signal comes directly from the output of IC609B.

4-12   VIDEO TERMINAL (KEYBOARD/CRT)

The video terminal can be used as an input/output device for microcomputers and many other digital electronic systems. Video terminals are particularly effective in inquiry/response situations where information is required for immediate use. Unless the CRT portion of the video terminal is attached to a printing device, no permanent record of displayed data is kept. As shown in Fig. 4-38, the video terminal is essentially a cathode-ray tube or CRT (complete with vertical and horizontal sweep circuits similar to those used in TV sets) plus a keyboard. Information is displayed on the face of the CRT immediately after the information is received from the microcomputer or entered from the keyboard.

Operation of the video terminal keyboard is essentially the same as for an electric typewriter described in Sec. 4-11. However, in a video terminal, the keyboard functions only as an input device. That is, the keyboard output is converted into data bytes (in machine language suitable for input to the microcomputer) by a decoder (which is generally part of the video terminal). The output function (display of data from the microcomputer) is performed by the CRT. Standard electromagnetic sweep deflection (similar to a TV set) is used for both the horizontal and vertical CRT deflection. The electron

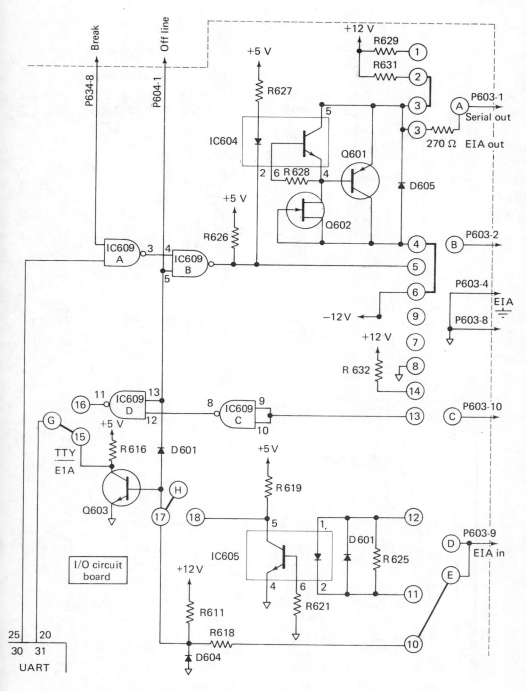

Figure 4-35 Typical EIA input/output circuit connections.

271

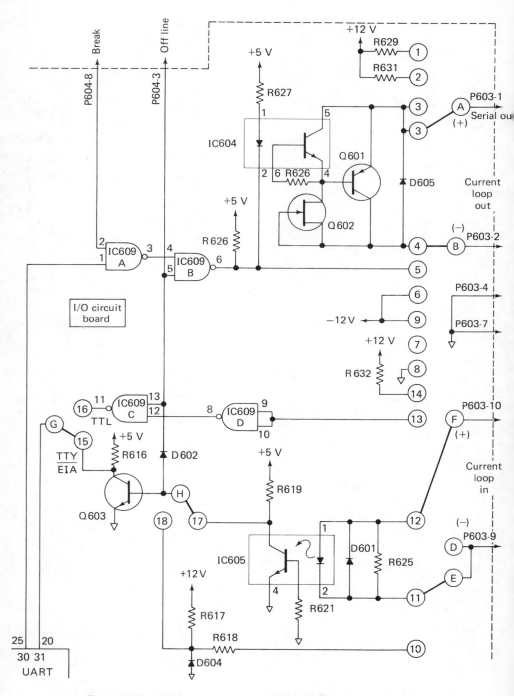

Figure 4-36 Typical 20-mA current loop input/output circuit connections.

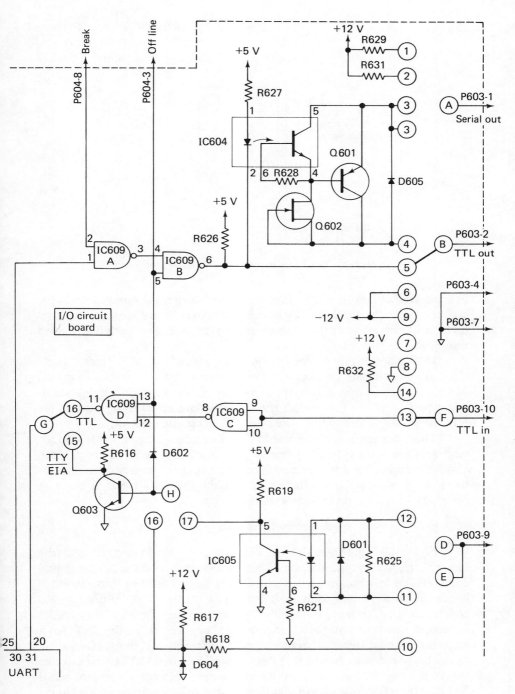

Figure 4-37 Typical TTL input/output circuit connections.

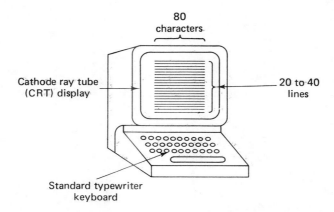

Figure 4-38 Typical video terminal (CRT display/keyboard).

beam is swept across the CRT face several hundred (or thousand) times per second. Typically, the deflecting circuits sweep out 20 to 40 rows or horizontal lines of characters (letters, numbers, or symbols), with each line divided into 80 spaces (for 80 characters).

In all video terminal systems, the persistence of the CRT screen is fairly short. This is necessary so that the readout is removed in a reasonable period of time (to permit new data bytes to be read in). Some video terminals contain storage registers that permit the displayed characters to be retained (or refreshed) on the screen until the operator clears the readout.

There are many systems used to produce the characters. Most video terminals use some type of *character generator,* usually in IC form. Such character generators are a form of decoder and convert binary data bytes into electrical voltages that alter or manipulate the CRT electron beam as necessary to trace out characters on the screen.

4-12-1   Typical video terminal circuits

Figure 4-39 is the overall block diagram of a typical video terminal (the Heathkit H9). This terminal can display information coming from a digital computer, or information typed in from the keyboard. The information is displayed on a 12-inch (diagonal) CRT that is capable of displaying 960 characters at one time (12 rows with 80 characters per row). A 67-key ASCII keyboard, which permits you to compose and edit directly on the CRT, has a capability of 64 different characters and 24 different functions. (Operation of the keyboard is similar to that described in Sec. 4-11-3.) The message or program from the keyboard can be transmitted through a standard serial I/O interface (Sec. 4-11-4) to a digital computer or through the parallel interface to a paper-tape reader/punch (Sec. 3-6-1).

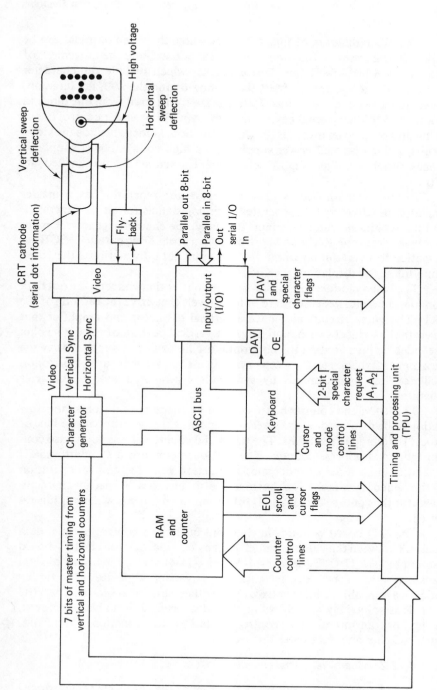

Figure 4-39 Video terminal block diagram.

275

The block diagram of Fig. 4-39 shows how the video terminal can be divided into a series of functional blocks that coordinate the movement of data on an 8-bit ASCII bus. These blocks, which include the character generator, the RAM and counter, the keyboard, and the I/O, are interconnected directly to the ASCII bus. Data bits are transferred from one block to another and to the external devices over the ASCII bus under control of the timing and processing unit (TPU). Other functional blocks include the video circuits, a line-operated power supply, and a high-voltage power supply or flyback system (similar to a TV set), which is operated from the video circuits.

As discussed in Sec. 4-12-2, the character generator has a master oscillator and dividers that generate the timing signals for all the functional modules within the video terminal. Basically, the character generator takes data bytes from the RAM and converts them from 8-bit parallel ASCII information to *serial dot information*. The serial dot information is then used to modulate the display on the CRT screen.

The video module contains a video amplifier that converts the dot data from the character generator into voltage levels that drive the cathode of the CRT. The video circuits also have a vertical oscillator and amplifier that generate the vertical sweep, and drive the vertical portion of the CRT yoke. The video circuits contain the horizontal oscillator and driver that drive the horizontal flyback and sweep systems. The vertical and horizontal oscillators are synchronized by sync pulses generated by the character generator.

The RAM and counter board contain the memory that stores the information to be displayed on the screen. Everything that appears on the face of the CRT is stored in the RAM. The RAM and counter circuit board also contain the RAM address counter. The keyboard contains a standard ASCII keyboard and a series of other special-function keys that control operation of the video terminal. The keyboard also includes a special character generator that puts selected ASCII information on the bus whenever it is required.

The I/O board contains latches and buffers to control parallel data transfers between the video terminal and any external device. The I/O board also contains a UART (Secs. 4-11-3 and 4-11-4) that controls serial data transfers between the video terminal and any external devices. The timing and processing unit (TPU) controls all the other functional blocks. The TPU is the decision-making module within the video terminal. Data bits are moved, written, or transmitted under control of the TPU as a function of the front-panel switches and data bytes themselves.

4-12-2   Character generator circuits

Figure 4-40 is an overall block diagram of the character generator circuits used in the video terminal. As discussed, the CRT screen is divided into

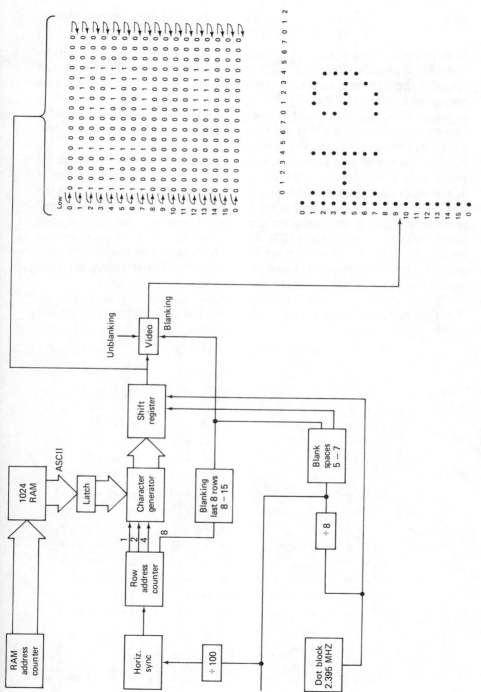

Figure 4-40 Character generator circuits, block diagram.

12 lines, each having 80 character locations, for a total of 960 character locations. Each location corresponds to a specific memory address in RAM. For instance, the character location at the upper-left corner of the screen might correspond to RAM address 00000000. A location somewhere in the middle of the screen might correspond to address 0111101011. Each character location on the screen is in a 128-space matrix arranged in 16 horizontal lines and 8 vertical rows as shown in Fig. 4-41.

At the end of a vertical retrace, the RAM address counter is initialized to the address for the upper-left corner of the screen. The address counter addresses one of the 960 RAM locations. Each memory location contains the ASCII code for the character to be displayed. The ASCII code is applied through a latch to the character generator. However, only the six least significant bits of the 7-bit ASCII word are used to address the character generator. Sixty-four characters are stored in the character generator ROM, each at its own address, as shown in Fig. 4-42(b). For instance, the 6-bit ASCII word 001000 addresses the character H.

Since the electron beam in the CRT scans only one line at a time, the dots that make up the characters must be shifted out serially, one row at a

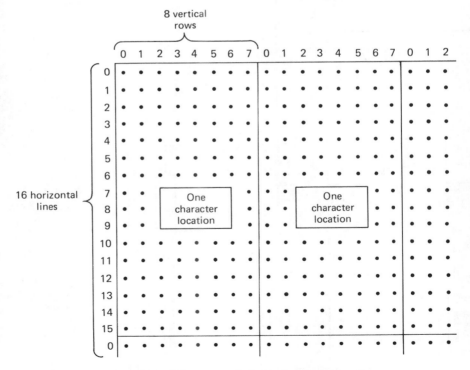

Figure 4-41 128-space matrix for each character location.

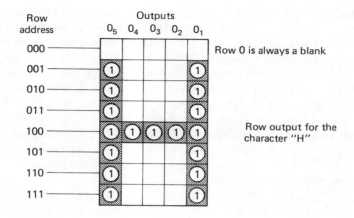

Character H

ASCII bit	6	5	4	3	2	1
Character generator input	A_9	A_8	A_7	A_6	A_5	A_4
6-bit ASCII word	0	0	1	0	0	0

(a)

Character address		A_9	0	0	0	0	1	1	1	1
		A_8	0	0	1	1	0	0	1	1
A_6	A_5	A_7 A_4	0	1	0	1	0	1	0	1
0	0	0								
0	0	1								
0	1	0								
0	1	1								
1	0	0								
1	0	1								
1	1	0								
1	1	1								

(b)

Figure 4-42 Arrangement of characters stored in character generator ROM.

time. To do this, the character generator puts out up to five dots in a row for each character. Figure 4-42(a) shows the row output for the character H. The 5-bit dot data bytes are always followed by three spaces (the absence of dots) before the dots for the next character are shifted out.

Since the CRT scanning process requires the dot data to be applied to the video circuits in a serial format, the 5-bit dot data bytes are loaded into a shift register and shifted eight times. This shifts out the 5 bits of the dot data, followed by three spaces. The three spaces are formed when a blanking signal is applied to the video circuits during the fifth through seventh shift sequence. The spaces are also necessary because the character location on the screen has been defined as being eight rows wide. The spaces provide space between characters.

During the fifth through the seventh shift sequence, the shift register is loaded with new dot data for the next character. The RAM address counter is also incremented to its next address. This process of addressing succeeding RAM locations and outputing dot data continues until a horizontal sync pulse is generated at the end of a scan line. The horizontal sync pulse increments the row counter, and causes the electron beam to begin a new scan line. The character generator now outputs dot data for the next row in the dot matrix. This process continues until all seven rows of data have been blanked. Horizontal scan lines 8 through 15 are normally blanked. A new character line (16 scan lines) begins after scan line 15 is completed.

4-13   OUTPUT PRINTERS

Where there is a limited amount of information to be read out of a microcomputer or other digital electronic device, the standard electric typewriter is the most convenient and in most common use. Such printers are sometimes known as console, message, or supervisory printers. Long rolls of paper are put into the typewriter carriage to record both the entry of data (by the operator at the keyboard) and the readout of processed data.

When high-speed readout is required or a large quantity of information is to be read out, various forms of printers are used. Note that many of these printers were used with electromechanical data-processing equipment long before computers or digital electronic equipment of any kind were invented. With a standard typewriter, only one letter can be typed at a time. Similarly, the up-and-down movement of type bars in a typewriter is far too slow for computers. Output printers overcome these problems.

There are two basic types of printers: (1) *line-at-a-time* printers, which print all characters on a given line simultaneously; and (2) *character-at-a-time printers,* which print each character serially, one position at a time, similar to the way a typewriter prints (but at much higher speeds).

No matter what printer is used, the printer receives signals from a decoder, or character generator, which converts data from the microcomputer into a form that is compatible with the printer. In some printers, a character generator similar to that described for the video terminal (Sec. 4-12) is used to convert machine language into characters. No matter what system is used, the microcomputer output buffer is synchronized with the output printer by *service request signals,* as discussed in Sec. 4-11-2. That is, the printer sends a service request signal to the microcomputer when it is ready to accept the next data byte (or character).

4-13-1 Line-at-a-time printers

Line-at-a-time printers include *bar* or *gang printers,* as well as *wheel, drum,* or *chain* printers, and *electrostatic* printers. As shown in Fig. 4-43, a drum printer uses a solid cylindrical drum, around which characters are embossed in a regular pattern of rows. The drum rotates at a constant speed. As the A row passes the line to be printed, hammers behind the paper strike the paper against the drum, causing one or more As to be printed. As the B row

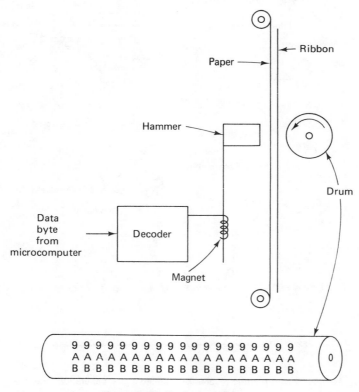

Figure 4-43 Basic drum printer.

moves into place, any print position requiring the letter B is printed in the same manner. One complete revolution of the drum is required to print each line.

The drum printer is often used with a commercial (point-of-sale) transaction terminal. The Seiko printer, shown in Fig. 4-44, is such a drum printer. As shown, the printer has a continually rotating print drum

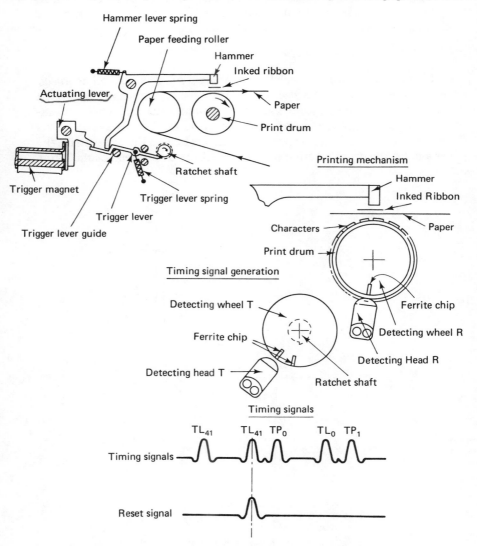

Figure 4-44 Seiko printing mechanism, timing signal generation, and timing signals.

mechanism which uses what is referred to as the *flying printer* technique. The print drum and the ratchet shaft are geared together and rotate continuously in the direction shown. During a nonprint condition, the right end of the trigger level is held away from the ratchet by the trigger lever spring. In the nonprinting condition, the trigger magnet is not actuated and the hammers are lifted upward to a neutral position by the hammer lever springs.

When actuated, the trigger magnet actuating lever forces the opposite end of the trigger lever against the ratchet. During its next rotation, the ratchet engages the right end of the trigger lever, causing a downward motion to the right-hand end of the hammer. The hammer thus strikes through the inked ribbon paper, causing the character then under the hammer to be printed.

Any of 42 characters (alphanumeric, plus special characters such as $, *, etc.) are printed in a 21-column format. Each column position has a complete character set spaced evenly around the drum. Because of a 42:1 gear ratio, the ratchet rotates 42 times for each complete drum rotation. Thus, each character of the set is positioned under a print hammer once during every rotation of the drum.

The decoders and control circuitry actuate the hammers at just the right time in response to timing signals generated electromagnetically by means of ferrite chips or magnets and detecting heads as shown. The timing signals or pulses are generated each time a ferrite magnet passes a detecting head. Rotation of the ratchet shaft generates signals TP and TL for each of the 42 characters. TP provides timing for energizing the trigger magnets, TL for deenergizing. A reset signal R is generated by each complete rotation of the drum. The resulting waveform for a complete drum rotation is shown in Fig. 4-44.

4-13-2 Character-at-a-time printers

The *teletype printer* prints one character at a time in a manner similar to that of an electric typewriter. However, all the type is placed on a square block rather than on individual hammers. The type block moves from left to right, positioning the proper character at each print position. As the block stops at each position, a hammer strikes the proper character from behind. The character is depressed against an inked ribbon which, in turn, imprints the character on the paper. Typical teletypeprinter speed is about eight lines per minute.

The *matrix printer* consists of pins placed in a 5×7 matrix, as shown in Fig. 4-45. The characters are formed when the appropriate pins strike against the paper. (In some matrix printers, the pins burn spots onto the paper.) The pins are selected to form characters by a character generator or decoder that operates in a manner similar to that for the video terminal (Sec. 4-12). Some matrix printers are capable of printing about 700 to 800 lines per minute, making the unit comparable in speed to most line-at-a-time printers.

4-14 DISKETTE (FLOPPY DISK) RECORDING BASICS

The floppy disk is the most commonly used magnetic storage device for microprocessor-based systems. The floppy disk itself (often referred to as a *diskette*) is a removable magnetic storage media which is permanently contained in a paper envelope. A *diskette drive* is a low-cost peripheral that performs the electromechanical and read/write functions necessary to record and recover data on the diskette. Figure 4-46 shows a typical diskette and diskette drive mechanism. The diskette is similar to the 45-rpm phonograph record. However, the diskette rotates at a speed of 360 rpm (typical), and the data bytes are stored magnetically on concentric tracks over the face of the diskette.

Data bytes are recorded serially on the diskette. Usually, because of the high serial data rates, it is necessary to use special circuits for the serial/parallel conversion, data recovery, and data error checking when interfacing a diskette and a microprocessor system. The hardware that performs this function is often called a *formatter*. The formatter also serves as a buffer between the microprocessor system and the diskette, as shown in Fig. 4-47. The formatter combines with the microprocessor system to control the diskette drive, and sometimes the combination is called a *floppy disk controller*. As used here, the term "controller" includes not only the system hardware, but also those microprocessor programs which directly or indirectly control the diskette drive. The program routines for the diskette are often referred to as floppy disk (or diskette) *drivers* or *control modules*.

4-14-1 Interfacing diskettes with microprocessors

There are several considerations that must be made when interfacing diskettes with microprocessors. We will not discuss all of them here, since

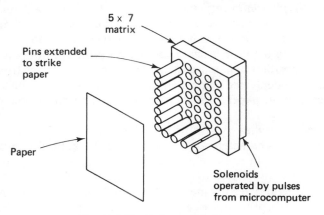

5 x 7 matrix

Pins extended to strike paper

Paper

Solenoids operated by pulses from microcomputer

Figure 4-45 Basic matrix printer.

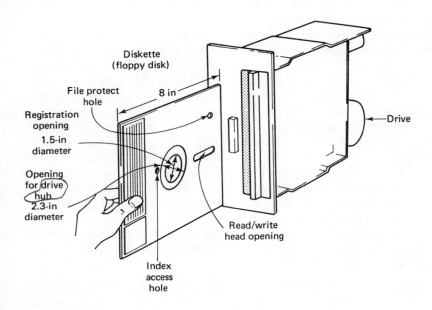

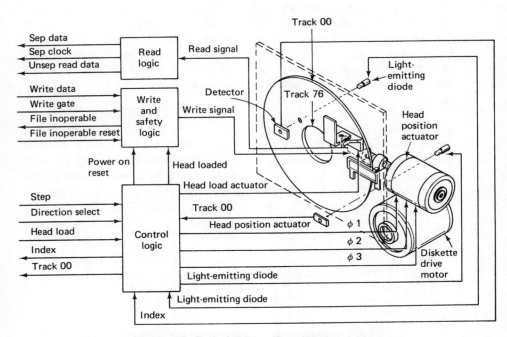

Figure 4-46 Typical diskette (floppy disk) and drive.

they are essentially the problem of the system designer. However, two considerations are worth mentioning.

Some diskette drives are designed for "daisy-chain" interfacing, where some of the interconnecting lines are shared and some are dedicated. Other drives use the "radial" interface, where all the interconnecting lines are dedicated. Both interfacing techniques are shown in Fig. 4-48. Each technique has its advantages. The radial interface isolates (or buffers) each drive, whereas the daisy-chain interface requires less system hardware.

No matter what interfacing system is used, the microprocessor is, in effect, busy all the time when used with a diskette. This is due to the high data rates involved. It also means that no other microprocessor peripherals can be serviced while in a diskette read or write operation. This is true provided that the transfer of data is controlled by the microprocessor and not via some type of direct memory access (DMA) hardware. Since no other peripherals can be serviced, interrupts generated by the other system elements must be disabled during diskette read or write operations. Allowances must be made in the system design to permit 100 percent system dedication to the diskette during read or write operations.

Direct memory access (DMA).   Many microcomputer systems use some form of DMA, which allows data to be transferred between a floppy disk (or other peripheral) and memory without interference to the microprocessor. DMA is often referred to as a method for speeding up data movement between elements of the microcomputer system. DMA allows fast peripherals (or perhaps another microprocessor) access to the system memory without taking up microprocessor time. In actual practice, however, DMA does use some microprocessor program time. How much time depends on the DMA system in use.

In a typical system, there are two DMA lines (for example, DMA-IN and DMA-OUT) provided for special types of byte transfer between memory and the peripheral. Activating the DMA-IN line causes an input byte to be immediately stored in a memory location without intervention by the program being executed. Activating the DMA-OUT line causes a byte to be immediately transferred from memory to the requesting peripheral. A register within the microprocessor is used as a DMA *pointer*. This built-in DMA

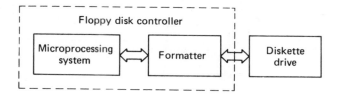

Figure 4-47   Relationship between microprocessor system and diskette.

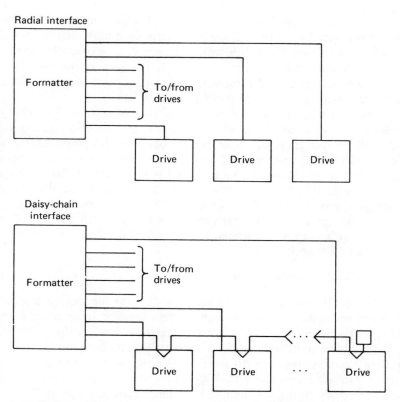

Figure 4-48 Basic arrangements for radial and daisy-chain interfacing formats.

memory pointer register is used to indicate the memory location for the DMA cycles. The program initially sets the DMA pointer to a beginning memory location. Each DMA byte transfer *automatically increments the pointer* to the next-higher memory location. Repeated activation of a DMA line can cause the transfer of any number of consecutive bytes to and from memory, independent of concurrent program execution.

4-14-2 Seek and restore operations

Information is transferred to and from the diskette by read/write heads in a manner similar to that described in Sec. 3-6-2 for magnetic tape recording. (However, data bytes are stored serially on a diskette, whereas most tape cassettes use parallel recording of the data bytes.) One of the functions of the diskette drive is to position the read/write head over the appropriate track where the data bytes are to be recorded or read. This function is sometimes referred to as a *seek and restore* operation.

The diskette records data on 77 circular tracks numbered 00 to 76. In order to access a certain record, the read/write head must first be locked in position at the track which contains that record. The operation that performs the head movement function is called a *seek operation*. For the diskette, a seek is executed by stopping the head one track at a time. The timing between steps is controlled from an interval timer.

The *restore operation* is similar to the seek operation. The main difference between seek and restore is that a restore operation always moves the read/write head to track 00. After the seek operation is completed, the only way to verify that the proper track has been accessed is to read the track address magnetically recorded on the track (in a manner similar to that for tape, as described in Sec. 3-6). When track 00 is accessed, the diskette drive usually generates a "track 00" status signal or flag. As shown in Fig. 4-46, the track 00 signal is often developed by a sensor (such as an LED and photocell). In some units, a similar sensor detects when the diskette is at an "index" or start position.

4-14-3 Data access time

One of the primary reasons to use a diskette for storage instead of another type of magnetic device, such as a tape cassette, is improved data access time. By definition, access time includes *seek time,* or the time for the read/write head positioner to move from its present location to the newly specified location (typically 10 ms/track); *settle time,* or the time for the positioner to settle onto a new track (typically 10 ms from the last step pulse); and *latency time,* or the time required for the diskette to rotate to the desired position (typically 83.3 ms, average). The diskette spins at a fixed rate of 167 ms per revolution. On the average, the data will be one-half of a revolution, or 83.3 ms, away from the head. This is known as *average latency time.*

4-14-4 Typical diskette specifications

A typical diskette system records and reads data at 250K bits/s or 4 μs per bit. When used with a typical 8-bit parallel microprocessor system, the data rate is $250 \div 8 = 31.25$K bytes/s, or 32 microseconds per byte. A single (typical) diskette has a capacity of 2,050,048 bits, or 256,256 bytes on the 77 tracks. There are 26 sectors on each track and 128 bytes on each sector. Track recording formats are discussed further in Sec. 4-14-6.

4-14-5 Functional description of a typical diskette drive

The diskette drive shown in Fig. 4-46 consists of read/write and control electronics, drive mechanism, read/write head, track positioning mechanism, and the removable diskette. These components perform the following function: interpret and generate control signals, move read/write head to the selected track, and read and write data. The relationship and interface signals for the internal functions of the drive are shown in Fig. 4-46.

General operation. The head-positioning actuator positions the read/write head to the desired track on the diskette. The head load actuator loads the diskette against the read/write head and data bits may then be recorded or read from the diskette. The head position actuator, which consists of an electrical stepping motor and lead screw, positions the lead screw clockwise or counterclockwise in 15° increments. A 15° rotation of the lead screw moves the read/write head one track position. The microprocessor system increments the stepping motor to the desired track. The diskette drive motor rotates the spindle at 360 rpm through a belt-drive system. Fifty- or 60-Hz power can be accommodated by changing the drive pully.

Signal interface. The signal interface between the diskette drive and the microprocessor system consists of lines required to control the drive and to transfer data to and from the drive. All lines in the signal interface are digital in nature (the signals are in pulse form) and either provide signals to the drive (input) or provide signals to the microprocessor (output).

Input signals. There are six input signal lines:

    1. *Direction select,* which defines the direction of motion of the read/write head when the *step* line is pulses. An open circuit or binary 1 defines the direction as out, and if a pulse is applied to the step line, the read/write head will move away from the center of the diskette (toward track 00). Conversely, if the direction select input is shorted to ground, or binary 0, the direction of motion is defined as in, and if a pulse is applied to the step line, the read/write head will move toward the center of the diskette (toward track 76).

    2. *Step input,* which is a control signal that causes the read/write head to move with the direction of motion defined by the direction select line. The access motion is initiated on each binary 0 to binary 1 transition of the step input signal.

    3. *Load head,* which is a control signal to an actuator that allows the diskette to be moved into contact with the read/write head. A binary 1 deactivates the head load actuator and causes a bail to lift the pressure pad from the diskette. This removes the load from the diskette and read/write head. A binary 0 level on the load head line activates the head load actuator and allows the pressure pad to bring the diskette into contact with the read/write head (with the proper contact pressure).

    4. *File inoperable reset,* which provides a direct reset for the file inoperable output signal. The file inoperable condition is reset when a binary 0 is applied to the file inoperable reset line.

5. *Write gate,* which controls the writing of data on the diskette. A binary 1 on the write gate line turns off the write function. A binary 0 enables the write function and disables the stepping circuitry.

6. *Write data,* which provides the data to be written on the diskette.

Output signals.   There are six output signal lines:

1. *Track 00,* which indicates when the read/write head is positioned at track zero (the outermost data track).

2. *File inoperable,* which is the output of the data safety circuitry and is a binary 0 level when a condition that jeopardizes data integrity has occured.

3. *Index,* which is a signal provided by the diskette drive once each revolution (166.67 ms) to indicate the beginning of the track. Normally, the index signal is at binary 1 and makes the transition to the 0 level for a period of 1.7 ms once each revolution.

4. *Separated data,* which comprises the interface line over which read data bits are sent to the microprocessor system. The signals written on the diskette (using a frequency-modulation system) are demodulated by the drive electronics and are converted to data pulses. The data pulses (representing the data bytes on the diskette) are sent over the separated data line. Normally, the separated data signal is at binary 1, and each data bit recorded on the diskette causes the signal to make the transition to a 0 level for 200 ns.

5. *Separated clock,* which provides the microprocessor system with the clock bits recorded on the diskette. The levels and timing are identical to the separated data line except that a separated clock pulse occurs each 4 μs. The recoding formats are discussed in Sec. 4-14-6.

6. *Unseparated read data,* which provides raw data (clock and data bits together) to the microprocessor.

4-14-6   Diskette recording formats

The format of the data recorded on the diskette is totally a function of the microprocessor system. Data bytes are recorded on the diskette using frequency modulation as the recording mode. Each data bit recorded has an associated clock bit recorded with it. Data bits written on and read back from the diskette take the form shown in Fig. 4-49. As shown, the clock bits and data bits (if present) are interleaved in the form of *bit cells.* By definition, a bit cell is the period between the leading edge of one clock bit and the leading edge of the next clock bit.

A *data byte,* when referring to serial data being written onto or read from the diskette drive, is defined as eight consecutive bit cells. The MSB cell is defined as bit cell 0 and the LSB cell is defined as bit cell 7. When reference is made to a specific data bit (such as data bit 3), it is with respect to the corresponding bit cell (bit cell 3).

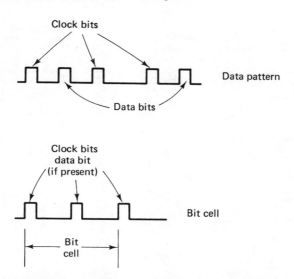

Figure 4-49 Basic diskette data bit recording format, showing formation of a bit cell.

During a write operation, bit cell 0 of each byte is transferred to the diskette drive first, with bit cell 7 being transferred last. Correspondingly, the most significant byte of data is transferred to the diskette first, and the least significant byte last. When data bytes are being read back from the drive, bit cell 0 of each byte is transferred first, with bit cell 7 last. As with reading, the most significant byte is transferred first from the drive to the microprocessor. Figure 4-50 illustrates the relationship of the bits within a byte. Figure 4-51 shows the relationship of bytes for read and write data.

Track formats. Each of the 77 tracks (00 through 76) may be formatted in numerous ways, dependent upon the microprocessor system. The diskette drive shown in Fig. 4-46 uses either *index recording* or *sector recording*.

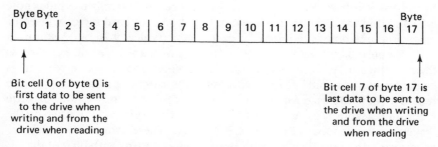

Figure 4-50 Relationship of bits within a data byte as recorded on a diskette (floppy disk).

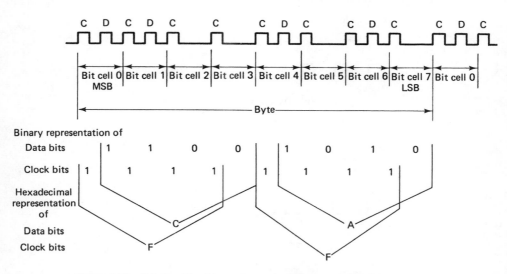

Figure 4-51 Relationship of bytes for read and write data as recorded on a diskette (floppy disk).

Index recording format. With index recording, the microprocessor may record one long record or several smaller records. Each track is started by a physical index pulse, and then each record is preceded by a unique recorded identifier. This type of recording is called *soft sectoring*. Figure 4-52 shows a typical index recording format.

Sector recording format. With sector recording, the microprocessor may record up to 32 sectors (or records) per track. (Some systems use 26 sectors per track.) Each track is started by a physical index pulse, and each sector is started by a physical sector pulse. This type of recording is called *hard sectoring*.

Keep in mind that the recording format is controlled primarily by the microprocessor. However, the choice between hard and soft sectoring is set by the diskette drive. For soft sectoring (index recording) only one index hole is required in the diskette, as shown in Fig. 4-46. For hard sectoring (sector recording) one hole is required for each sector (typically 26 or 32 holes).

Because of the great variety of formats available, we will not discuss them further here. However, it should be noted that diskette recordings generally include some feature to ensure the safety and accuracy of the records. As is the case for tape recordings, described in Sec. 3-6, diskette recordings generally use a form of CRC. Each field (Fig. 4-52) written on the diskette is appended with two CRC bytes. These two CRC bytes are generated by circuits that count all the data bits, starting with bit 0 of the address mark and ending with bit 7 of the last byte within a field (excluding the

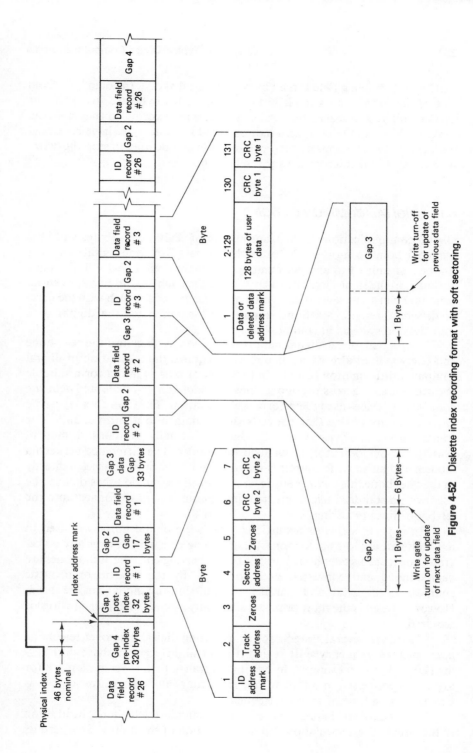

Figure 4-52 Diskette index recording format with soft sectoring.

CRC bytes). When a field is read back from the diskette, the data bits (from bit 0 of the address mark to bit 7 of the second CRC byte) are divided by the CRC number. A nonzero remainder indicates an error bit in the data read back, while a remainder of zero indicates that the data bytes have been read back correctly. The nonzero (error) indication can be used to stop the drive, send a flag to the microprocessor, and so on.

4-15 DIGITAL COMMUNICATIONS

Digital communications is the transmission of digital data (bytes and bits) from one point to another. For example, the data from one digital computer can be transmitted to another computer or other digital device in the same building, in the same city, or to a city across the country. Similarly, a central computer can serve several users on a *time-sharing* basis, each with the users at different remote locations. Figure 4-53 summarizes the digital data transmission systems in common use.

At present, most digital data transmission systems use the telephone and teletype lines already available. This requires the translation of digital equipment information (usually in the form of pulses) into a form suitable for transmission across telephone lines. The *data set, modem,* and *acoustic coupler* are devices used to perform the translation. The term data set can be applied to any device that converts digital data into a form suitable for transmission, and vice versa. A data set usually contains a modem (*mo*dulator-*dem*odulator) or an acoustic coupler. The difference between a modem and an acoustic coupler is that the modem must be connected directly to the telephone lines. An acoustic coupler produces audio tones that can be used with a standard telephone, making it unnecessary to connect into the telephone lines or wiring.

In most modems or acoustic couplers, the digital pulses are used to modulate an audio tone (or tones) that can be transmitted over any voice-grade line (such as telephone lines). At the receiving end, the audio tones are demodulated and converted back to pulses by the modem or acoustic coupler. Both serial and parallel transmission methods can be used. However, serial is the most popular, since only a single line (pair of wires) is required.

There are several systems for transmitting digital data over telephone lines, such as frequency shift keying (FSK) and differential phase shift keying (DPSK). As an example, in one digital transmission system, a mark (or logic 1) is converted to a tone of one frequency, and a space (or logic 0) is converted to a tone of another frequency.

Many data sets in present use include a standard telephone, in addition to the circuits for modulation and demodulation of the pulses. This permits

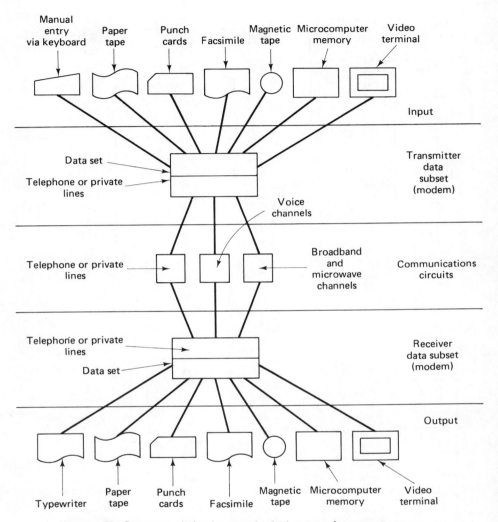

Figure 4-53   Data transmission (communication) systems in common use.

operators to communicate before and after the transmission of digital data. In a typical situation, the operator will dial a number, putting the operator in touch with operators at the opposite end of the line. After making the necessary arrangements, data can be transmitted back and forth between the locations via the modem or acoustic coupler, and the lines.

Any of the input/output units described in this chapter can be used as terminal devices for digital communications, in association with the data set. The video terminal described in Sec. 4-12 is one of the most popular terminal

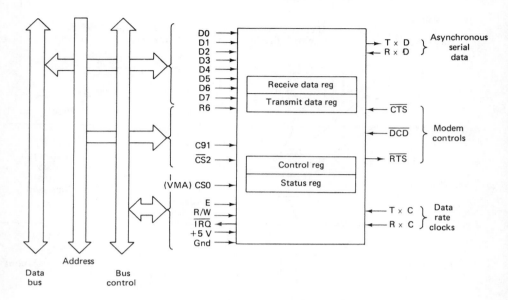

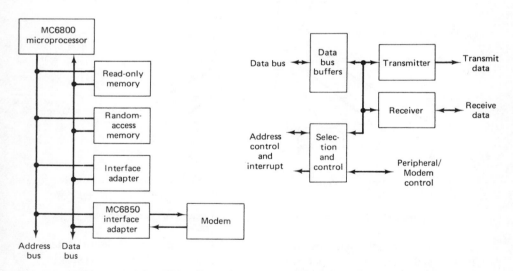

Figure 4-54 MC6850 ACIA, interfacing, relationship to system, and basic block diagram.

devices. For example, a time-sharing user can have a visual display terminal and a data set at an office location connected by telephone lines to a computer at some remote point. The input and output of the video terminal is connected through the data set to the telephone lines. With arrangement, the video terminal is (in effect) connected directly to the computer.

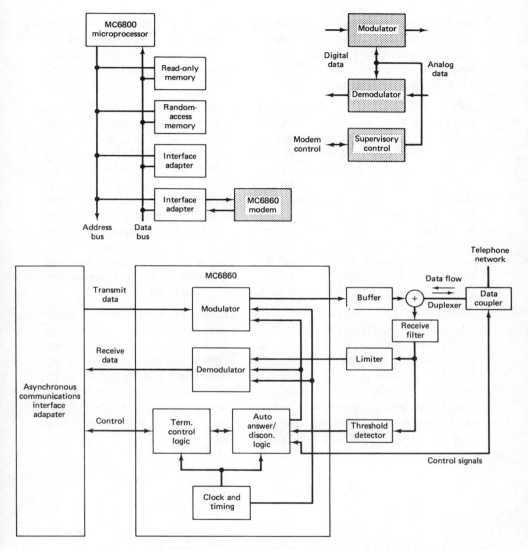

Figure 4-55 MC6860 digital modem, relationship to system, basic block diagram, and typical system configuration.

4-15-1 Typical serial data transmission system

The following paragraphs provide a brief description of a typical serial data transmission system, which is used with the Motorola 6800 microprocessor system (discussed in Sec. 4-4 and shown in Fig. 4-12). Two devices are involved, the MC6850 ACIA (asynchronous communications interface adapter) and the MC6860 digital modem. Keep in mind that these are only two of many interface and data transmission devices available for use with the Motorola 6800. Other microprocessor manufacturers produce many similar interface and data transmission devices. Both the MC6850 and MC6860 are in IC form.

ACIA.   Figure 4-54 shows the interfacing relationship to the system and basic block diagram of the ACIA. As shown, the ACIA provides the data formatting and control to interface asynchronous serial data between the microprocessor and modem. The parallel data of the M6800 bus system is serially transmitted and received (full-duplex) by the ACIA, with proper formatting and error checking. (The term "full-duplex" applied here means that serial data can be transmitted and received automatically on the same lines.) The functional configuration of the ACIA is programmed via the data bus during system initialization. A programmable control register provides variable word lengths, clock division ratios, transmit control, receive control, and interrupt control. Three I/O lines are provided to control external peripherals or modems. A status register is available to the microprocessor and reflects the current status of the transmitter and receiver.

Digital modem.   Figure 4-55 shows the relationship to the system, the block diagram, and a typical system interface for the digital modem. The modem is a subsystem designed to be integrated into a wide range of equipment (not just the M6800) using serial data communications. The modem provides the necessary modulation, demodulation, and supervisory control functions to implement a serial data communications link, over a voice-grade channel (telephone line), using FSK at bit rates up to 600 bits per second. The modem can be implemented into a wide range of data storage devices, remote data communications terminals, and I/O interfaces.

CHAPTER

5

Testing
and Troubleshooting
Digital Equipment

This chapter describes the procedures and equipment required for test and troubleshooting of digital electronic equipment. In many cases, the test equipment and techniques used for other solid-state electronic devices can be carried over into digital electronics. In other cases, special tools and techniques are required for even simple digital equipment. In this chapter, we concentrate on basic approaches to digital troubleshooting. These approaches can be applied to any digital electronic system now being developed, to those that will be manufactured in the future, and to digital equipment now in use. For a thorough discussion of digital electronic troubleshooting, your attention is invited to the author's best selling *Handbook of Practical Microcomputer Troubleshooting* (Reston, Va.: Reston Publishing Company, Inc., 1979).

Although the classic theoretical and practical principles of electronic test and troubleshooting apply to digital equipment, there are many practical techniques that apply only to digital. As one example, the key technique described here and in most digital literature is based on measurement of pulses at inputs and outputs of digital circuits. From a troubleshooting standpoint, a complete digital logic diagram can appear as a hopeless maze, particularly if the digital circuits are *interlaced* (that is, multiple functions for the same signal at different times, or one signal depending on many other signals or signal conditions). However, if pulses can be checked at the input and/or output of each circuit or group of circuits, the condition of that circuit can be checked quickly.

The troubleshooting technique requires the use of an oscilloscope and possibly a pulse generator. Such a combination of test equipment will quickly determine the presence or absence of pulses at circuit inputs and outputs, as well as the pulse duration, amplitude, and delay. For example, an AND gate may show an incorrect output even though there are two input pulses of correct amplitude and duration, if one pulse is delayed. These measuring and monitoring techniques, together with many other test and fault localization methods for digital circuits, are discussed throughout this chapter.

5-1   DIGITAL TEST EQUIPMENT

Most test and troubleshooting procedures for digital equipment can be performed using conventional meters (including high-voltage meters for measurement of video terminal CRT voltages), multitrace oscilloscopes (for measurement of pulses on data and address buses, clock and other control lines, etc.), and assorted clips, patchcords, power supplies, and handtools. Theoretically, all digital electronic troubleshooting problems can be solved using such instruments. However, there are some specialized test instruments that greatly simplify digital equipment service. Such specialized equipment includes pulse generators, probes, pulsers, current tracers, logic clips, comparators, and logic analyzers. We shall concentrate on such specialized test equipment in this section. The remaining sections of this chapter describe test and troubleshooting approaches using the equipment covered here.

It is not the purpose of this chapter to promote one type of test and troubleshooting equipment over another (or one manufacturer over another). Instead, the chapter is devoted to the basic operating principles or features of test/troubleshooting instrument types in common use. You may then select the type of digital test equipment best suited to your own needs and pocketbook.

Although complicated theory has been avoided, the following discussions cover the way each type of equipment is used in digital test and troubleshooting, and what signals or characteristics are to be expected from each. We also describe how the features and outputs found on present-day test equipment relate to specific problems in digital service. Specific techniques for using the equipment in troubleshooting situations are discussed in the remaining sections of this chapter.

A thorough study of this chapter will familiarize you with the basic principles and operating procedures for typical equipment used in general digital troubleshooting. It is assumed that you will take the time to become equally familiar with the principles and operating controls for any particular test/service equipment that you use. Such information is contained in the user's manuals for the particular equipment. It is absolutely essential that you become thoroughly familiar with your own particular equipment. No amount of textbook instruction will make you an expert in operating test equipment; it takes actual practice.

It is strongly recommended that you establish a routine operating procedure, or sequence of operation, for each item of test/troubleshooting equipment. This approach will save time and familiarize you with the capabilities and limitations of your own equipment, thus minimizing the possiblity of false conclusions based on unknown operating conditions.

5-1-1   Safety precautions in digital test/troubleshooting

In addition to a routine operating procedure, certain precautions must be observed during operation of any electronic test equipment. Many of these precautions are the same for all types of test equipment; others are unique to special test instruments, such as meters, oscilloscopes, and probes. Some of the precautions are designed to prevent damage to the test equipment, or to the circuit where the troubleshooting operating is being performed. Others are meant to prevent injury to you. Where applicable, special safety precautions are included throughout this chapter. The following general safety precautions should be studied thoroughly and then compared to any specific precautions called for in the test equipment literature.

1.   The most hazardous elements in digital troubleshooting are the high voltages used for the CRT of video terminals. Typically, the anode of a CRT requires about 1000 V/in. Thus, a 12-inch CRT requires about 12,000 V for the anode. It is essential that the anode voltage (and the related power supply) be measured with a properly shielded, *high-voltage probe* similar to those used in television service, as described in the author's best selling *Handbook of Simplified TV Service* (Englewood Cliffs, N.J.: Prentice-Hall, Inc., 1978).

2.   The next most hazardous problem arises from the fact that most digital equipment operates from a 115-V line power. This voltage is sufficient to cause death. Even though the voltage is usually dropped to about 5

or 6 V for use by the ICs, the power-supply circuits still operate at the line voltage.

3.   Many troubleshooting instruments are housed in metal cases. These cases are connected to ground of the internal circuit. For proper operation, the ground terminal of the instrument should always be connected to ground of the digital device being serviced. Make certain that the chassis or housing of the digital device being serviced is not connected to either side of the power line or to any potential above ground. If there is any doubt, connect the digital device being serviced to the power line through an isolation transformer.

4.   Remember that there is always danger in servicing digital devices that operate at hazardous voltages (such as the high voltage used for the CRT), especially as you pull off the digital equipment cover and reconnect the power cord. Always make some effort to familiarize yourself with the digital device before servicing it, bearing in mind that high voltages may appear at *unexpected points in defective equipment.*

5.   It is good practice to remove power before connecting test leads to high-voltage points. It is preferable to make all service connections with the power removed. If this is impractical, be especially careful to avoid accidental contact with circuits and objects that are grounded. Keep in mind that even low-voltage circuits may present a problem. For example, a screwdriver dropped across a 5-V line can cause enough current to burn out a major portion of the circuits, possibly beyond repair. Of course, this problem is nothing compared to the possible injury to yourself. Working with one hand away from the equipment and standing on a properly insulated floor lessen the danger of electrical shock.

6.   Capacitors may store a charge large enough to be hazardous. Discharge filter capacitors before attaching test leads.

7.   Remember that leads with broken insulation offer the additional hazard of high voltages appearing at exposed points along the leads. Check test leads for frayed or broken insulation before working with them.

8.   To lessen the danger of accidental shock, disconnect test leads immediately after the test is completed.

9.   Remember that the risk of severe shock is only one of the possible hazards. Even a minor shock or burn can place you in danger of more serious risks, such as a bad fall or contact with a higher voltage source.

10.   The experienced service technician is continuously on guard against injury and does not work on hazardous circuits unless another person is available to assist in case of accident.

11.   Even if you have considerable experience with the test equipment used in troubleshooting, always study the service literature of any instrument with which you are not thoroughly familiar.

12. Use only shielded leads and probes. Never allow your fingers to slip down to the meter probe tip when the probe is in contact with a "hot" circuit.

13. Avoid vibration and mechanical shock. Most electronic test equipment is delicate.

14. Study the circuit being serviced before making any test connections. Try to match the capabilities of the instrument to the circuit being serviced.

5-1-2 Oscilloscopes for digital work

An oscilloscope for digital work should have a bandwidth of at least 50 MHz and preferably 100 MHz. The pulse width or pulse durations found in most digital equipment are on the order of a few microseconds, often only a few nanoseconds wide.

A *triggered horizontal sweep* is an essential feature. Preferably, the delay introduced by the trigger should be very short. Often, the pulse to be monitored occurs shortly after an available trigger. In other cases, the horizontal sweep must be triggered by the pulse to be monitored. In some oscilloscopes, a delay is introduced between the input and the vertical deflection circuits. This permits the horizontal sweep to be triggered before the vertical signal is applied, thus assuring that the complete pulse will be displayed.

Dual-trace or multiple-trace sweeps are also essential. Most digital troubleshooting is based on the monitoring of two time-related pulses (say, an input pulse and output pulse, or a clock and a readout pulse). With dual trace, the two pulses can be observed simultaneously, one above the other or superimposed if convenient. In other test configurations, a clock pulse is used to trigger both horizontal sweeps. This allows for a three-way time relationship measurement (clock pulse and two circuit pulses, such as one input and one output). A few oscilloscopes have multiple-trace capabilities. However, this is not standard. Usually, such an oscilloscope is provided with plug-in options that increase the number of horizontal sweeps.

The sensivity of both the vertical and horizontal channels should be such that full-scale deflection can be obtained (without overdriving or distortion) with less than a 1-V signal applied. Typically, the signal pulses used in digital work are on the order of 5 V, but often they are less than 1 V (such as when ECL is involved).

In some cases, *storage oscilloscopes* (for display of transient pulses) and *sampling oscilloscopes* (for display and measurement of very short pulses) may be required. However, most digital work can be done with an oscilloscope having the bandwidth and sweep capabilities just described.

5-1-3 Multimeters for digital work

The voltmeters for digital work should have essentially the same characteristics as for other solid-stage troubleshooting. As usual, meters with digital readouts ae easier to use than meters with analog (moving-needle) readouts. Generally, a very high input impedance is not critical for digital work (200,000 Ω/V is usually sufficient).

Typically, digital circuit operating voltages are 12 V, while the logic voltages (pulse levels) are 5 V or less. Thus, the voltmeter should have *good resolution on the low-voltage scales*. Here again, the meter with a digital readout is an advantage. For example, a typical logic level might be where 0 is represented by 0 V, and 1 is represented by 3 V. This means that an input of 3 V or greater to an OR gate will produce a 1 (or true) condition, while an input of something less than 3 V, say 2 V, will produce a 0 (false) condition. As is typical for digital logic specifications, the region between 2 V and 3 V is not spelled out. If the voltmeter is not capable of readout clearly between 2 and 3 V on some scale, a false conclusion can easily be reached. If the OR gate in question is tested by applying a supposed 3 V, which was actually 2.7 V, the OR gate might or might not operate to indicate the desired 1 (or true) output.

Generally, d-c voltage accuracy should be ± 2 percent or better, with ± 3 to ± 5 percent accuracy for the a-c scales. The a-c scales will probably be used only in checking power-supply functions in strictly digital equipment, since all signals are in the form of pulses (requiring an oscilloscope for display and measurement).

The ohmmeter portion of the multimeter should have the usual high-resistance ranges. Many of the troubles in the most sophisticated and complex computers boil down to such common problems as cold solder joints, breaks in printed circuit wiring that result in high resistance, as well as shorts or partial shorts between wiring (producing an undesired high-resistance condition). The internal battery voltage of the ohmmeter should not exceed any of the voltages used in the circuit being checked.

5-1-4 Pulse generators for digital work

Ideally, a pulse generator should be capable of duplicating any pulse present in the circuits being tested. Thus, the pulse generator output should be continuously variable (or at least adjustable by steps) in amplitude, pulse duration (or width), and frequency (or repetition rate) over the same range as the circuit pulses. That is not always possible in every case. However, with modern laboratory pulse generators, it is generally practical for most digital equipment. Typically, the pulses are ± 5 V or less in amplitude, but could be 10 to 15 V in rare cases (such as when CMOS is operated with a high voltage supply). The pulses are rarely longer than 1 second or shorter than 1 ns, although there are exceptions here, too. Repetition rates are generally less than 100 kHz but could run up to 10 MHz.

Some pulse generators have special features, such as two output pulses with variable delay between the pulses or an output pulse that can be triggered from an external source. However, most routine digital test and troubleshooting can be performed with standard pulse generators.

5-1-5   Logic probe

Figure 5-1 shows a typical logic probe. Such probes will detect and indicate high and low (1 or 0) logic levels, as well as intermediate or "bad" logic levels, including an open circuit, on a single line of a digital circuit. For example, the probe can be connected to one of the address or data bus lines (or the clock line, chip-select line, etc.) and will indicate the state (0, 1, or high-impedance) of that line. Note the extender clamp connected to an IC at the left of the probe in Fig. 5-1. Such extenders are used where the IC terminals are not readily accessible for connection of test leads.

Figure 5-1   Hewlett-Packard 545A logic probe.

A logic-level indicator lamp, near the probe tip, gives an immediate indication of the logic states, either static or dynamic, existing in the circuit under test when the probe is touched to the circuit or line. Figure 5-2 shows how the logic-level indicator lamp responds to voltage levels and pulses found on a typical TTL digital circuit line. The indicator lamp can give any of four indications: (1) off, (2) dim (about half brilliance), (3) bright (full brilliance), or (4) flashing on and off. The lamp is normally in its dim state, and must be driven to one of the other three states by voltage levels at the probe tip. The lamp is bright for inputs at or above the 1 state and off for inputs at or below 0. The lamp is dim for voltages between the 1 and 0 states and for open circuits. Pulsating inputs cause the lamp to flash at about a 10-Hz rate.

The logic probe is particularly effective when used with the logic pulser described in Sec. 5-1-6. For example, the logic pulser can be used to simulate the input to an element in the digital circuit (such as the clock input to a microprocessor, RAM, ROM, or I/O), while the logic probe is touching the output of the element to sense activity of the element.

Using the logic probe in digital circuits. The following paragraphs provide brief descriptions of how the logic probe can be used for troubleshooting. More detailed procedures and approaches are given in the remaining sections of this chapter.

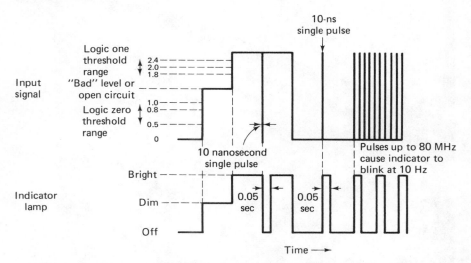

Figure 5-2 How the logic-level indicator lamp responds to voltage levels and pulses found on a typical TTL digital circuit line.

Probe power supply. The logic probe can be powered from the digital equipment power supply or from a regulated d-c power supply. If a separate power supply is used, the power supply and digital equipment grounds should be connected together. The power-supply voltage range for TTL operation is 4.5 to 15 V. A ground wire (provided with the probe) may be connected just behind the probe indicator window. The ground wire is a convenient means of connecting grounds when using external regulated power supplies. The ground wire also improves pulse-width sensitivity and noise immunity. However, use of the ground wire is optional and is not required for all applications.

Pulse detection. The logic probe is ideal for detecting short-duration and low-repetition-rate pulses that would be difficult to observe on an oscilloscope. Typically, positive pulses of about 10 ns or greater will trigger the indicator on for at least 50 ms. Negative pulses cause the indicator to go off momentarily.

Testing three-state logic outputs. The bad-level feature of the logic probe is useful for testing three-state logic outputs. The logic-high and logic-low states are detected as described under pulse detection. The third state found in many microcomputer circuits (the high-impedance output) is detected as an open-circuit (or bad-level condition), which leaves the probe indicator dim. The bad-level indication is also useful for detecting floating or unconnected TTL inputs, which look like a bad level to the logic probe.

Basic probe techniques. Several logic circuit analysis techniques are useful with the logic probe, as discussed in remaining sections of this chapter. One technique is to run the digital device under test at its normal clock rate while monitoring for various control signals, such as reset, halt, memory read, flag, or clock. Questions such as "is there a clock signal" or "is there a memory read signal" are quickly resolved by noting if the probe indicator is flashing on and off, indicating that pulse train activity is present. For example, if there is no flashing indication when the probe is connected to the clock line, there is no clock pulse. Or if the probe indicator does not flash when connected to a memory read line in a microcomputer, the microprocessor is probably stuck somewhere in the program (since the memory read line is usually pulsed at each step of the program to open the RAM or ROM at each address).

Another useful technique is to replace the normal clock signal with a very slow clock signal from a pulse generator, or to *single step* the clock input with a pulse generator similar to the logic pulser described in Sec. 5-1-6. This allows changes in logic signals to occur at a rate slow enough that they can be observed on a *real-time* basis. (The basic single-stepping technique is discussed in Sec. 5-1-10.) Real-time analysis, coupled with the ability to inject logic level pulses anywhere in the microcomputer with a logic pulser and the means to detect logic state changes with the logic probe, contributes to rapid troubleshooting and fault finding in the control lines. Unfortunately, logic probes are not as valuable with trying to locate faults in multiple-line circuits, such as the data and address buses.

5-1-6 Logic pulser

Figure 5-3 shows a typical logic pulser being used with a current tracer (Sec. 5-1-7). Such pulsers are hand-held logic generators used for injecting controlled pulses into digital logic circuits such as microcomputers. The electronics are housed in a hand-held probe. Automatic pulse control is provided for TTL and many other logic families. Thus, the logic pulses are compatible with most digital devices. Pulse amplitude depends on the logic supply voltage (3 to 18 V), which is also the supply voltage for the logic pulser.

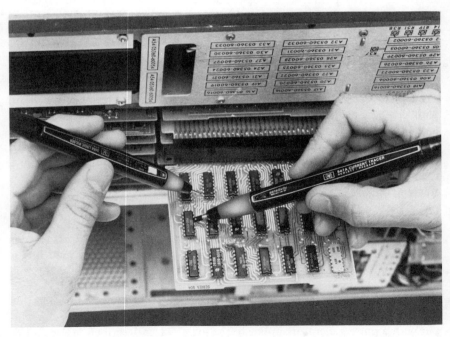

Figure 5-3 Hewlett-Packard 546A logic pulser (left) shown with Hewlett-Packard 547A current tracer.

Pulse current and pulse width depend on the load being pulsed. The frequency and number of pulses generated by the pulser are controlled by operation of a push-slide switch on the pulser probe. A flashing LED indicator located on the tip indicates the output mode.

The logic pulser generates pulses as programmed by the fingertip push-slide switch. The pulser is programmed by pushing the switch once for each single-pulse output, or a specific number of times for continuous pulse streams or pulse bursts at selected frequencies. The pulses are also applied to an LED indicator on the pulser tip. (The pulses applied to the LED are slowed down for visibility.)

Pressing the switch automatically drives an IC output or input from low to high (0 to 1) or from high to low (1 to 0). The high source and sink current capability of the pulser can override IC output points, originally in either the 0 or 1 state. The nominal 10-μs pulse width is long enough for even slow CMOS circuits to accept, but heavy circuit loads (such as TTL drivers) result in narrower pulses that limit the amount of energy delivered to the device under test.

The pulser output is three-state. In the off state, the pulser's high output impedance ensures that circuit operation is unaffected by probing until the pulser switch is pressed. Pulses can be injected while the circuit is operating, and no disconnects are needed. Some probes have a multipin kit or accessory that provides up to four pulses simultaneously (say to eliminate 4 bits of an 8-bit word on a data or address bus).

Using the logic pulser in digital circuits.   The following paragraphs provide brief descriptions of how the logic pulser can be used for troubleshooting. More detailed procedures and approaches are given in the remaining sections of this chapter. Logic pulsers are generally used with other instruments, such as with the logic probe (Sec. 5-1-5), current tracer (Sec. 5-1-7), logic clip (Sec. 5-1-8), and logic comparator (Sec. 5-1-9), for routine digital electronic troubleshooting.

Logic gate testing.   A logic gate may be tested by pulsing the gate's input while monitoring the output with a logic probe, as shown in Fig. 5-4. The logic pulser generates a pulse opposite to the state of the input line and can change the gate output state. This assumes that the output of the gate is not clamped in its state by another input (such as a high on the other input of an OR gate).

If the pulse is not detected at the output, pulse the output line as shown in Fig. 5-5. If the output is not shorted to V_{CC} or common, the logic probe should indicate a pulse opposite to its original indication. If not, check for external shorts (solder bridges and the like) before removing the IC. Shorts

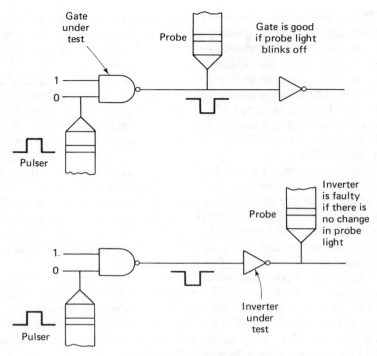

Figure 5-4 Logic gate testing with probe and pulser.

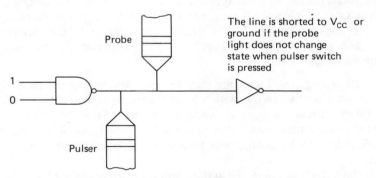

Figure 5-5 Testing for output shorts with pulser and probe.

at the inputs and outputs are best located by means of the current tracer described in Sec. 5-1-7.

Capacitor pulse generator. When a logic pulser is not available, some technicians use bus voltages to trigger circuits manually. That is, the pulse trains normally present on a particular circuit line or at a particular input are removed and replaced by pulses manually injected one at a time by momentarily connecting the line or input to a bus of appropriate voltage and polarity. This is not recommended, for two reasons: (1) some circuits can be damaged by prolonged application of the voltage; and (2) in other circuits, no damage will occur, but the results will be inconclusive. For example, many faulty circuits will operate normally when a bus voltage is applied (even momentarily) but will not respond to a pulse.

A better method is to use a capacitor as a pulse generator. The basic technique is shown in Fig. 5-6. Simply charge the capacitor by connecting it between ground and a logic bus (typically 5 or 6 V). Then connect the charged capacitor between ground and the input to be tested. The capacitor will discharge, creating an input pulse. Be sure to charge the capacitor to the correct voltage level and polarity. Generally, the capacitor value is not critical. Try a 1-μF capacitor as a starting value. Often, it is convenient to connect on lead of the capacitor to a ground clip, with the other lead connected to a test prod. The capacitor can be clipped to ground, and then the prod tip can be moved from bus to input, as needed.

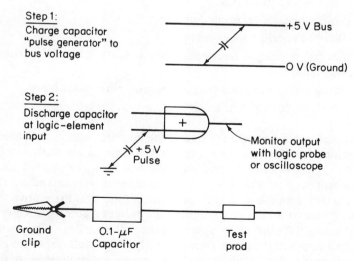

Figure 5-6 Capacitor pulse generator for single-pulse testing of digital circuits.

5-1-7 Current tracer

Figure 5-3 shows a current tracer being used with a typical logic pulser (Sec. 5-1-6). Such current tracers are hand-held probes that enable the precise localization of low-impedance faults in many electrical systems (including typical digital electronic printed-circuit wiring). The current tracer senses the magnetic field generated by a pulsing current internal to the circuit or by current pulses supplied by an external simulus such as the logic pulser. Indications of the presence of current pulses is provided by the lighting of the indicator lamp near the current tracer tip. Adjustment of current tracer sensitivity over the 1-mA to 1-A range is provided by a sensitivity control near the indicator. The current tracer is self-contained and requires less than 75 mA at 4.5 to 18 V from any convenient source.

Using the current tracer in digital circuits. The following paragraphs provide brief descriptions of how the logic pulser can be used for troubleshooting. More detailed procedures and approaches are given in the remaining sections of this chapter.

The current tracer operates on the principle that whatever is driving a low-impedance fault node or point must be delivering the majority of the current. Tracing the path of this current leads directly to the fault. Problems that are compatible with this method are the following:

1. Shorted inputs of ICs.
2. Solder bridges on printed-circuit boards.
3. Shorted conductors in cables.
4. Shorts in voltage distribution networks (such as V_{CC}-to-ground shorts).
5. Stuck data or address buses, including three-state buses.
6. Stuck wired-AND structures.

Basic current tracer operation. Use of the current tracer is indicated when conventional troubleshooting reveals a low-impedance fault (such as a short). In typical operation, you align the mark on the current tracer tip along the length of the printed-circuit trace at the driver, and adjust the sensitivity control until the indicator lamp just lights. You then move the current tracer along the trace (line) or place the current tracer tip directly on the terminal points (IC pins) while observing the indicator light. This method of following the path of the current leads directly to the fault responsible for the abnormal current flow. If the driving point does not provide pulse stimulation, the terminal may be driven externally by using a logic pulser at the driving point. The following paragraphs describe troubleshooting techniques for some of the more common problems.

Wired-AND problems. One of the most difficult problems in troubleshooting ICs of any type is a stuck wired-AND circuit (Sec. 2-8-1). Typically, one of the open-collector gates connected in the wired-AND mode may still continue sinking current after the gate has been turned off. The current tracer provides an easy method of identifying the faulty gate. (Of course, if the gate is located in an IC, the entire IC must be replaced. However, you have still located the problem and can take whatever action is best suited to the situation.)

Referring to Fig. 5-7, place the current tracer on the gate side of the pull-up resistor. Align the mark on the tracer tip along the length of the printed-circuit trace (line) and adjust the tracer's sensitivity control until the indicator is just fully lighted. If the indicator will not light, use a logic pulser to excite the line as shown. Place the tracer tip on the output pin of each gate; only the faulty gate (or pin) will cause the indicator to light.

Gate-to-gate faults. When a low-impedance fault (short, full, or partial) exists between two gates, the current tracer and logic pulser combine to quickly pinpoint the defect. In Fig. 5-8, the output of gate A is shorted to ground. Place the pulser midway between the two gates, and place the cur-

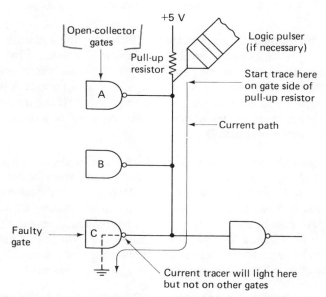

Figure 5-7 Locating wired-AND problems with a current tracer.

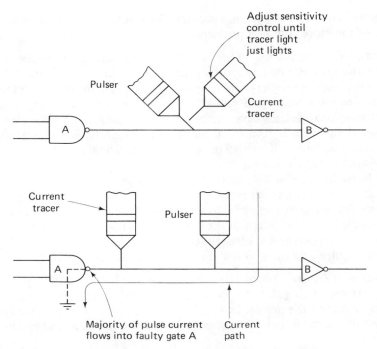

Figure 5-8 Locating gate-to-gate faults with a current tracer.

rent tracer tip on the pulser pin as shown. Pulse the line and adjust the current tracer sensitivity control until the indicator just lights. First place the current tracer tip next to gate A, and then gate B, while continuing to pulse the line. The tracer will light only on the gate A side, since gate A (the defect in this example) is sinking the majority of the current. If the tracer does not light when placed between the pulser and gate A, look for a short on the line between the pulser and gate B.

Solder bridge/cable problems. When checking printed-circuit traces (lines) that may be shorted by solder bridges or other means, start the current tracer at the driver and follow the trace. Figure 5-9 shows an example of an incorrect path due to a solder bridge. As the tracer probe follows the trace from gate A toward gate B, the indicator remains lighted until the tracer passes the bridge. This is an indication that current has found some path other than the trace. Visually inspect this area for solder splashes and the like. These principles also apply when troubleshooting shorted cable assemblies.

Multiple gate inputs. Another type of IC arrangement is the one-output, multiple-input configuration. Figure 5-10 shows this type of circuit being

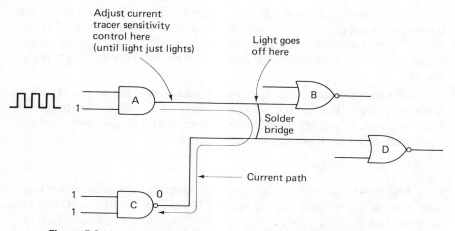

Figure 5-9 Locating solder bridges in printed-circuit lines with a current tracer.

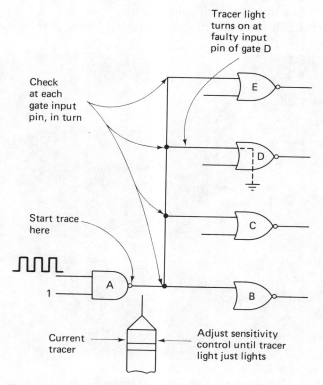

Figure 5-10 Checking multiple gate inputs with a current tracer.

pulsed by a signal on the input of gate A. In this case, place the current tracer tip on the output pin of gate A and adjust the sensitivity control until the indicator light just comes on. Then check the input pins of gates B through E. If one of the input pins is shorted, that pin will be the only one to light the indicator.

Should the tracer fail to light when placed next to the output of gate A, it is a good indication that the problem exists in gate A. To be sure that this is true, use the current tracer in the manner described under gate-to-gate faults. If the circuit has no input signal, use a logic pulser.

5-1-8 Logic clip

Figure 5-11 shows a typical logic clip. Such clips are designed for logic-level determination only on ICs using TTL and DTL circuits. Generally, clips can test flip-flops, gates, counters, buffers, adders, shift registers, and the like, but will not test ICs with nonstandard input levels or expandable gates. Clips will instantly and continuously show the logic levels (0 or 1) at all pins of a dual-in-line IC. All 16 input pins are electrically buffered to minimize loading on any circuit being tested.

Sixteen LEDs are the high and low (0 or 1) logic-level indicators. No

Figure 5-11 Hewlett-Packard 548A logic clip.

power-supply connections need be made, since the clip powers itself from the circuit under test by automatically locating the V_{CC} and ground pins of the IC. In use, you squeeze the thick end of the clip to spread the contacts, and place the clip on the IC to be tested. The LEDs on top of the clip will indicate the logic levels at each connected IC pin. The clip may be turned in either direction. Each of the clip's 16 LEDs independently follows level changes at the associated input. A lighted LED corresponds to a logic 1.

The real value of the logic clip is the ease of use. The clip has no controls to be set, needs no power connections, and requires practically no explanation as to how it is used. Since the clip has its own gating logic for locating the ground and 5-V pins, the clip works equally well upside down or right side up. Buffered inputs ensure that the circuit under test will not be loaded down. Simply clipping the unit onto a TTL or DTL IC package of any type makes all logic states visible at a glance.

The logic clip is much easier to use than either an oscilloscope or a voltmeter when you are interested in whether a lead is in the 1 or 0 state, rather than in the lead's actual voltage. The clip, in effect, is 16 binary voltmeters, and you do not have to shift your eyes away from the circuit to make the readings. The fact that a lighted LED corresponds to a logic 1 greatly simplifies the troubleshooting procedure. You are free to concentrate your attention on the circuits rather than on measurement techniques.

When the clip is used on a real-time basis (when the clock is slowed to about 1 Hz or is manually triggered) timing relationships become especially apparent. The malfunction of gates, FFs, counters, and adders then becomes readily visible as all of the inputs and outputs of an IC are seen in perspective. When pulses are involved, the logic clip is best used with the logic probe. Timing pulses can be observed on the probe, while the associated logic-state changes can be observed on the clip.

Figure 5-12 is the block diagram of a typical logic clip. As shown, each pin of the logic clip is internally connected to a decision gate network, a threshold detector, and a driver amplifier connected to an LED. Figure 5-13 shows the decision sequence of the decision gate network. In brief, the decision gate networks do the following:

1. Find the IC V_{CC} pin (power voltage) and connect it to the clip power voltage bus. This also activates an LED.
2. Find all logic high pins and active corresponding LEDs.
3. Find all open circuits and activate corresponding LEDs.
4. Find the IC ground pin, connecting it to the clip ground bus, and blank the corresponding LED.

The threshold detector measures the input voltage. If the voltage is not over the threshold voltage, the LED is not activated. An amplifier at the out-

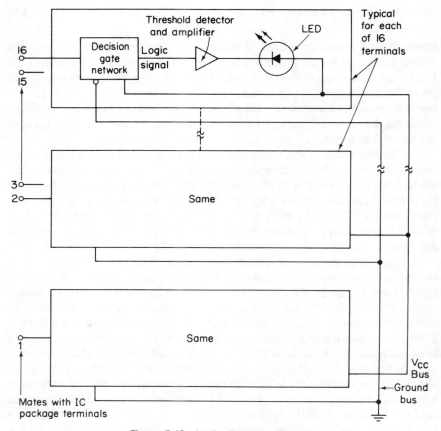

Figure 5-12 Logic clip block diagram.

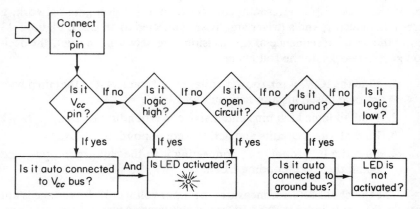

Figure 5-13 Logic clip decision sequence.

put of the threshold detector drives the LED. The LED will indicate high (will glow) if the IC pin is above 2 V and will indicate low (no glow) if the pin is below 0.8 V. If the IC pin is open, the LED will also show a high.

Using the logic clip in digital circuits.   At present, logic clips will accommodate only 16-pin ICs. Thus, logic clips are of little value when checking digital devices such as microprocessors, where a typical microprocessor has 40 pins, and 24 pins is standard for most ROMs, RAMs, and I/Os. However, there may be 40- and 24-pin logic clips in the future. Also, in some microcomputers, there may be a 16-pin IC that has access to the data and address buses.

If a logic clip can have access to both buses, the clip can be used for a *program trace* function, using *single stepping,* as described in Sec. 5-1-10. In such a case, the clip serves to read out the data word at each address of the program. However, at this writing, the *logic analyzer* is the most practical instrument for making a rapid, thorough program trace.

5-1-9 Logic comparator

Figure 5-14 shows a typical logic comparator. Such comparators clip onto 16-pin ICs and, through a comparison scheme, instantly display any logic-state differences between the test IC and a reference IC. Logic differences are identified to the specific pin or pins of the IC with the comparator's display of 16 LEDs. A lighted LED corresponds to a logic difference.

In use, the IC to be tested is first identified. A reference board with a good IC of the same type is then inserted in the comparator. The comparator is clipped onto the IC in question, and an immediate indication is given if the test IC operates differently from the reference IC. Even very brief dynamic errors are detected, stretched, and displayed.

The comparator operates by connecting the test and reference IC inputs in parallel. Thus, the reference IC sees the same signals that are inputs to the test IC. The outputs of the two ICs are compared, and any difference in outputs greater than 200 nS in duration indicate a failure. A failure on an input pin, such as an internal short, appears as a failure on the IC driving the failed IC. Thus, a failure indication actually pinpoints the malfunctioning pin.

As in the case of the logic clip (Sec. 5-1-8), logic comparators are presently limited to 16-pin ICs. Thus, both instruments are of little value in present-day microcomputers. However, 40- and 24-pin comparators may be developed in the future.

5-1-10 Logic analyzer

The logic analyzer is the most useful instrument for test and troubleshooting of any programmed digital equipment. Before going into

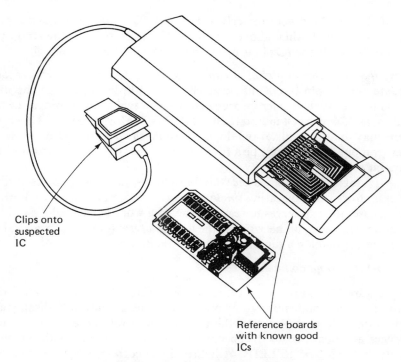

Clips onto
suspected
IC

Reference boards
with known good
ICs

Figure 5-14   Hewlett-Packard 10529A logic comparator.

how logic analyzers are used with programmed systems, let us discuss the
need for a logic analyzer (also known as a logic state analyzer).

The classic approach for troubleshooting any programmed device is to
monitor a significant system function (such as the data and address buses),
go through each step in the program, and compare the results with the program listing for each address and step. This is sometimes known as a *program trace* or, more simply, a *trace* function. One technique for this procedure is called *single stepping*. With the single-stepping approach, you
remove the normal clock pulses and replace them with single one-at-a-time
pulses obtained from a switch or pushbutton. This permits you to examine
and compare the data at each address with that shown in the program.

For example, assume that each of the eight lines on a data bus is connected to a multitrace oscilloscope so that the bit on each line appears as a
pulse on a corresponding trace. A pulse on the oscilloscope trace indicates a
binary 1, whereas the absence of a pulse indicates a binary 0 on that line. A
simplified version of such a test setup is shown in Fig. 5-15.

The test is started by applying a single pulse to the reset or clear line.
then sufficient single pulses are applied to the clock line until address number

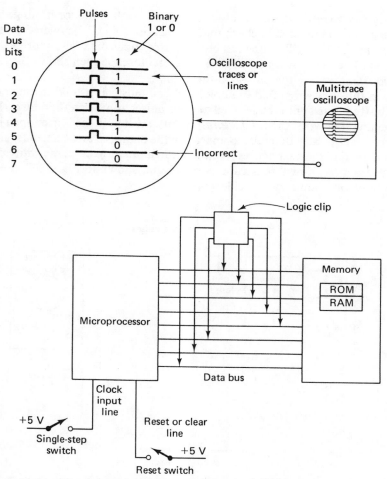

Figure 5-15 Basic microcomputer troubleshooting approach for timing problems.

1 (0001) appears on the address bus. Now assume that the program listing shows that a hex 7F (binary 0111-1111) data byte should be on the data bus at address 0001, but that the oscilloscope traces show binary 0011-1111. Obviously, the wrong instruction will be applied to the microprocessor and the system will malfunction. This could be caused by a broken line on the data bus, by a defect in the memory, by the absence of a memory-read pulse or a memory-read pulse that appears at the wrong time (opens the memory data buffer too soon or too late), or by several other possible causes. However, you have isolated the problem and determined where it occurs in the program.

Timing problems. If the problem appears to be one of timing, the oscilloscope can be used to check the time relationship of the related pulses. For example, the oscilloscope can be connected to the data bus, address bus, and read line, as shown in Fig. 5-16. The oscilloscope then shows the *time relationships* among the pulses on these lines.

In this very simplified example, the read pulse must hold the memory data buffer closed until the selected address pulses appear on the address bus (sometimes known as the *valid address* point), must hold the buffer open just long enough for all 8 data bits to appear on the data bus, and then must close the buffer until the next address is applied. In a practical case, the entire timing diagram (such as discussed in Sec. 4-8 and shown in Fig. 4-27) can be duplicated on a multitrace oscilloscope.

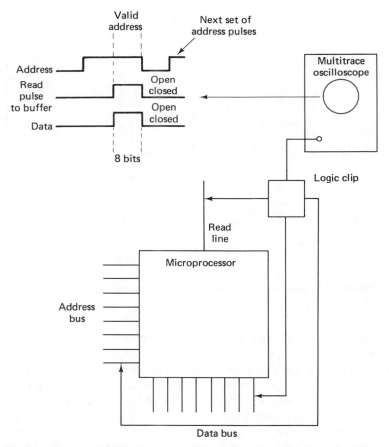

Figure 5-16 Basic microprocessor troubleshooting approach for timing problems.

Logic analyzer displays.   While single stepping and a check of system timing can pinpoint many digital equipment problems, there is an obvious drawback. Typically, a data byte is 8 bits and thus requires eight clock pulses or eight one-at-a-time pushes of the single-step button. Since all program steps require at least one byte (and often two or three bytes, possible 24 bits), you must, for example, push that button many times if the malfunction occurs at step 0333 of the program. This means that you must spend endless hours comparing program listings against binary readouts at addresses. (If you are already familiar with the troubleshooting of programmed devices, you know that the most time-consuming part of the task is in making such comparisons.)

The logic analyzer overcomes this basic problem by permitting you to select for display the data at a particular address. The logic analyzer will then run through the program at near the normal system speed (a fraction of a second) and display the selected data between the desirable *breakpoints* (points in the program before and after the area of interest). Figure 5-17 shows a typical logic analyzer and some related displays. A logic analyzer is essentially a multitrace oscilloscope combined with electronic circuits to produce special displays. The electronic circuits that produce these displays are sometimes known as *formatters.* The logic analyzer shown in Fig. 5-17 is operated by a keyboard. Other logic analyzers use switches and controls. No matter what control system is used, the logic analyzer is capable of three basic displays.

Timing display.   The multitrace feature of the logic analyzer can be used to reproduce *timing diagrams,* in a manner similar to that of a conventional multitrace oscilloscope. Figure 5-18 shows a typical timing diagram display as produced on a logic analyzer. (This is sometimes known as a time-domain display.)

Tabular data display.   The data display format of the logic analyzer (sometimes called the *data domain* format or display) is used to display data bytes (as they appear on the data and/or address buses) in binary form. In effect, the bytes appear as they would on paper (1s and 0s rather than the presence or absence of pulses). Figure 5-19 shows that several data bytes can be displayed simultaneously. This makes it possible to check the data words before and after the selected area of interest in the program. With some logic analyzers, the selected data word is indicated by extra brightness of the display. In Fig. 5-19, the top data word is the selected word (the breakpoint starts at this word and continues for 16 steps or words).

To use the tabular data format, you connect the logic analyzer to the data and/or address buses by means of patch cords (called logic probes) supplied with the analyzer. Then you select a particular data word and/or address breakpoint from the program listing by means of the logic analyzer

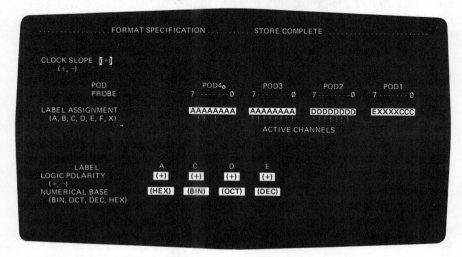

```
                        TRACE LIST                    TRACE COMPLETE

        LABEL       A     B     C     D     E     F       TIME
        BASE       HEX   HEX   HEX   OCT   BIN   DEC       DEC
                                                          (FEL)
        SEQUENCE    01    A     A     7     10    1
        START       01    A     A     7     10    0        21.8    US
          +01       01    B     0     2     11    1         3.7    US
          +02       01    B     1     6     10    1         3.5    US
          +03       01    B     2     3     10    1         4.7    US
          +04       01    B     3     0     00    1         3.5    US
          +05       01    B     4     7     10    1         3.5    US
          +06       01    B     5     2     01    1         4.7    US
          +07       01    B     6     6     10    1         3.5    US
          +08       01    B     7     3     01    1         4.7    US
          +09       01    B     8     0     00    1         3.5    US
          +10       00    7     D     0     00    1         3.5    US
          +11       00    7     E     0     00    1         4.7    US
          +12       00    7     F     0     00    1         4.7    US
          +13       00    8     0     0     01    1         4.7    US
          +14       00    8     1     6     10    1         5.8    US
          +15       00    8     2     0     00    1         3.6    US
          +16       00    D     2     7     01    1        10.6    US
          +17       00    D     3     1     10    1        12.5    US
          +18       00    D     4     1     10    1         4.7    US
```

```
              FORMAT SPECIFICATION              STORE COMPLETE


    CLOCK SLOPE  [+]
         (+, –)

              POD                 POD4₀       POD3        POD2        POD1
              PROBE             7 ······ 0   7 ······ 0  7 ······ 0  7 ······ 0

    LABEL ASSIGNMENT           AAAAAAAA    AAAAAAAA    DDDDDDDD    EXXXXCCC
    (A, B, C, D, E, F, X)
                                           ACTIVE CHANNELS


            LABEL              A         C         D         E
    LOGIC POLARITY            (+)       (+)       (+)       (+)
         (+, –)
    NUMERICAL BASE           (HEX)     (BIN)     (OCT)     (DEC)
    (BIN, OCT, DEC, HEX)
```

Figure 5-17 Hewlett-Packard 1610A logic analyzer and some related displays.

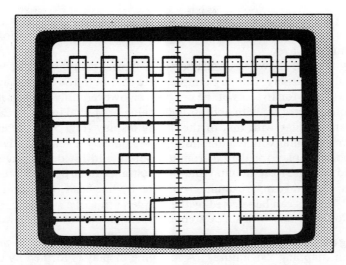

Figure 5-18 Typical timing diagram (time-domain) display as produced on a logic analyzer.

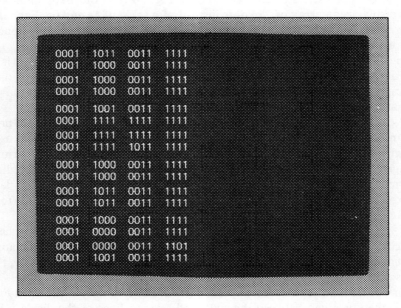

Figure 5-19 Typical tabular data (data-domain) display as produced on a logic analyzer.

controls (keyboard or switches) and start the program. The digital system then runs through the complete program, but only the desired portion of the program is displayed.

Figure 5-20 shows a comparison of a typical program listing versus a logic analyzer display. In this example, note that both the address and data bytes are shown for 16 steps of the program. However, only the first nine addresses are of interest, with address 0004 being of special interest. Compare this tabular display with that of the timing display (Fig. 5-18). Even those readers not familiar with troubleshooting will quickly realize the advantages of a tabular display for digital troubleshooting and for debugging of programmed devices. Keep in mind that the program can be operated at near its normal speed, and can be examined on a line-by-line basis, 16 lines at a time. The next 16 lines can be selected by the simple setting of a switch or touch of a key. Compare this with single stepping through the entire program!

Mapping display. Another display unique to logic analyzers is the mapping display or mapping "signature." A mapping display is formed by connecting the MSB bits of a data word to the vertical deflection circuits of the logic analyzer (those circuits that cause vertical deflection of the oscilloscope trace), while the LSB bits are connected to the horizontal deflection circuits. This produces a series of dots as shown in Fig. 5-21. In the mapping mode, the display is an array of 256 dots instead of a table of 1s and 0s. Each dot represents one possible combination of the 16 input lines so that any input is represented by an illuminated dot. An input of all 0s (00 00 in hex) is at the upper left corner of the display, whereas an input of all 1s (FF FF in hex) is at the lower right. The dots are interconnected so that the sequence of data changes can be observed. The interconnecting line gets brighter as it moves toward a new point, thereby showing the direction of data flow.

When the mapping display is used to monitor a digital equipment program or a portion of a program, the display assumes a unique pattern or "signature." Once you have learned to recognize the patterns for various programs, it is relatively easy to tell at a glance if the program is proceeding normally. Compare the table and map display of Fig. 5-22. The same data bytes (16 words, 16 bits per word) are in tabular form in Fig. 5-22(a) (top) and plotted in a map format in Fig. 5-22(b) (bottom).

Some logic analyzers are provided with a *cursor* function to help locate specific points in the mapping display. Generally, the cursor is a bright circle that can be manipulated over the display by the analyzer controls. In use, the cursor is positioned over an area of interest on the map, and the address or data word is read out on the analyzer controls. For example, as shown in Fig. 5-23, there is a gap between the sixth and seventh lines of the display. By positioning the cursor over the last dot in the sixth line and reading the corresponding word on the analyzer controls, you know the exact address or data word at which the malfunction occurs.

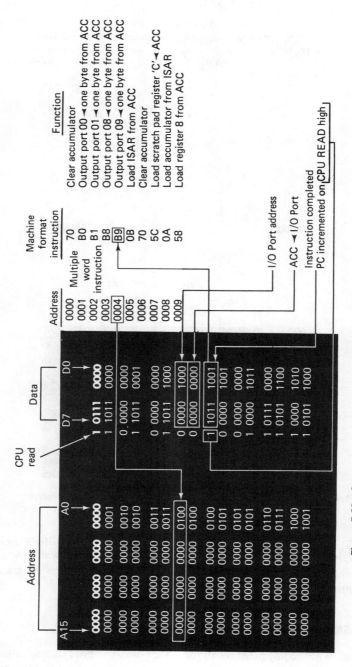

Figure 5-20 Comparison of a typical program listing versus a logic analyzer tabular data display.

Address	Machine format instruction	Function
0000	Multiple 70	Clear accumulator
0001	word B0	Output port 00 ◄ one byte from ACC
0002	instruction B1	Output port 01 ◄ one byte from ACC
0003	B8	Output port 08 ◄ one byte from ACC
0004	B9	Output port 09 ◄ one byte from ACC
0005	0B	Load ISAR from ACC
0006	70	Clear accumulator
0007	5C	Load scratch pad register 'C' ◄ ACC
0008	0A	Load accumulator from ISAR
0009	58	Load register 8 from ACC

I/O Port address

ACC ◄ I/O Port

Instruction completed

PC incremented on CPU READ high

327

Map display
(Positive logic, hexadecimal notation)

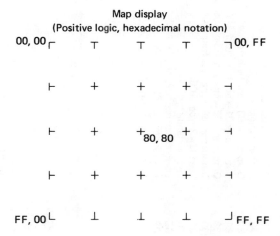

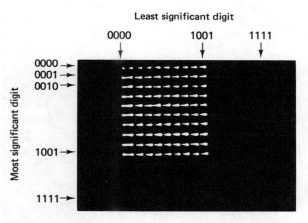

Figure 5-21 Typical mapping display as produced on a logic analyzer.

Keep in mind that the examples given here are very elementary. Logic analyzers, in the hands of experienced operators, are very powerful tools for testing, debugging, troubleshooting, and analyzing all forms of programmed digital equipment.

5-2 BASIC DIGITAL MEASUREMENTS

No matter what equipment is involved or what digital troubleshooting technique is used, you must be able to work with pulses. That is, you must be able to monitor pulses on an oscilloscope, measure their amplitude, duration

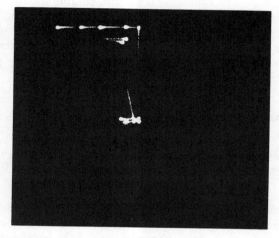

0010	0110	0001	1010
0000	0000	0001	1010
0000	0000	0001	0000
0000	0000	0000	0000
0000	0101	0001	1010
0010	0101	0001	1101
0010	0101	0001	1111
0010	0101	0001	1000
0010	0101	0001	1111
0000	0000	0001	1111
0000	0000	0001	0001
0000	0000	0000	0000
0000	0110	0001	1001
0010	0110	0001	1101
0010	0110	0001	1111
0010	0110	0001	1101

Figure 5-22 Comparison of a tabular display and mapping display as produced on a logic analyzer.

or width, and frequency or repetition rate. You must also be able to measure delay between pulses and to check operation of basic digital circuits or building blocks, such as OR gates, AND gates, FFs, delays, and the like. These basic digital measuring techniques, together with descriptions of the test equipment involved, are discussed in the following paragraphs.

5-2-1   Basic pulse measurement techniques

The following measurement techniques apply to all types of pulses, including square waves used in digital circuits.

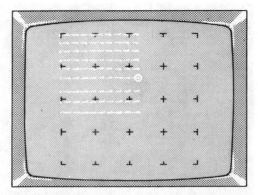

Figure 5-23   Using the cursor function of a logic analyzer mapping display to locate a malfunction in a program.

Pulse definitions.   The following terms are commonly used in describing pulses found in digital circuits. The terms are illustrated in Fig. 5-24. The input pulse represents an ideal waveform for comparison purposes. The other waveforms in Fig. 5-24 represent the shape of pulses that may appear in digital circuits, particularly after they passed through many gates, delays, and so on. The terms are defined as follows:

Rise time T_R: the time interval during which the amplitude of the output voltage changes from 10 percent to 90 percent of the rising portion of the pulse.

Fall time T_F: the time interval during which the amplitude of the output voltages changes from 90 percent to 10 percent of the falling portion of the pulse.

Time delay T_D: the time interval between the beginning of the input pulse (time zero) and the time when the rising portion of the output pulse attains an arbitrary amplitude of 10 percent above the baseline.

Storage time T_S: the time interval between the end of the input pulse (trailing edge) and the time when the falling portion of the output pulse drops to an arbitrary amplitude of 90 percent from the baseline.

Pulse width (or pulse duration) T_W: the time duration of the pulse measured between two 50 percent amplitude levels of the rising and falling portions of the waveform.

Tilt: a measure of the tilt of the full-amplitude, flat-top portion of a pulse. The tilt measurement is usually expressed as a percentage of the amplitude of the rising portion of the pulse.

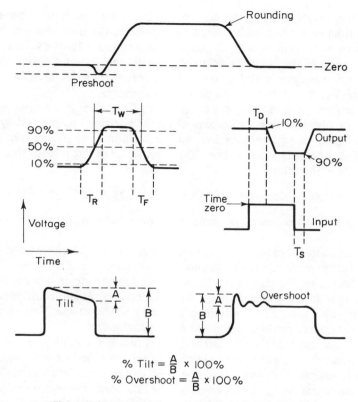

Figure 5-24   Basic pulse and square-wave definitions.

Overshoot or preshoot: a measure of the overshoot or preshoot occurring generally above or below the 100 percent amplitude level. These
measurements are also expressed as a percentage of the pulse rise.

These definitions are for guide purposes only. When pulses are very ir-
regular (such as excessive tilt, overshoot, etc.), the definitions may become
ambiguous.

5-2-2   Guidelines for rise-time measurements

Since rise-time measurements are of special importance in any pulse
circuit testing, the relationship between the oscilloscope rise time and the rise
times of the digital circuit components under test must be taken into account. Obviously, the accuracy of rise-time measurements can be no greater
than the rise time of the oscilloscope. Also, if the device is tested by means of
an external pulse from a pulse generator, the rise time of the pulse generator
must also be taken into account.

For example, if an oscilloscope with a 20-ns rise time is used to measure the rise time of a 15-ns digital circuit component, the measurement will be hopelessly inaccurate. If a 20-ns pulse generator and a 15-ns oscilloscope is used to measure the rise time of a component, the fastest rise time for accurate measurement will be something greater than 20 ns. Two basic guidelines can be applied to rise-time measurements.

The first method is known as the *root of the sum of the squares*. The method involves finding the squares of all the rise times associated with the test, adding these squares together, and then finding the square root of the sum. For example, using the 20-ns pulse generator and the 15-ns oscilloscope, the calculation is

$$20 \times 20 = 400 \qquad 15 \times 15 = 225 \qquad 400 + 225 = 625 \qquad \sqrt{625} = 25 \text{ ns}$$

This means that the fastest possible rise time capable of measurement is 25 ns.

One major drawback to this rule is that the coaxial or shielded cables required to interconnect the test equipment are subject to *skin effect*. As frequency increases, the signals tend to travel on the outside or skin of the conductor. This decreases conductor area and increases resistance. In turn, this increases cable loss. The losses of cables do not add properly to apply the root-sum-squares method, except as an approximation.

The second guideline or method states that if the equipment or signal being measured has a rise time *10 times slower* than the test equipment, the error is 1 percent. This amount is small and can be considered as negligible. If the equipment being measured has a rise time *three times slower* than the test equipment, the error is slightly less than 6 percent. By keeping these relationships in mind, the results can be interpreted intelligently.

5-2-3 Measuring pulse amplitude

Pulse amplitudes are measured on an oscilloscope, preferably a laboratory-type oscilloscope where the vertical scale is calibrated directly in a specific deflection factor (such as volts per centimeter). Most laboratory oscilloscopes have a step attenuator (for the vertical amplifier) where each step is related to a specific deflection factor. Typically, the pulses used in digital circuits are 5 V or less. Often, the pulse amplitude is critical to the operation of digital equipment. For example, an AND gate may require two pulses of 3 V to produce an output. If the two pulses are slightly less than 3 V, the AND gate may act as if there is not input or that both inputs are false. Thus, when troubleshooting any digital circuit, always check the actual pulse amplitude against that shown in the service literature. Generally, digital circuits are designed with considerable margin in pulse amplitude, but equipment aging or some malfunction can reduce the safety margin.

The basic test connections and corresponding oscilloscope display for pulse amplitude measurement are shown in Fig. 5-25. The basic procedure is as follows:

1. Connect the equipment as shown in Fig. 5-25. Place the oscilloscope in operation.

2. Set the vertical step attenuator to a deflection factor that will allow the expected signal to be displayed without overdriving the vertical amplifier. Typically, the vertical deflection factor should be 10 V or less full scale.

3. Set the input selector to measure ac. Pulses can also be measured with the oscilloscope set to dc. However, the trace will be moved up or down on the screen, depending on the polarity of the d-c voltage present with the pulse.

4. Connect the oscilloscope probe to the pulse signal being measured. Set the oscilloscope controls to display at least one and preferably two or more pulses on the screen. In most cases, this involves switching on the internal recurrent sweep such that the sweep is driven by the pulse train being measured. If the pulse being measured is of a one-shot nature where there is a relatively long time between pulses, it may be necessary to use an oscilloscope where there is a delay introduced between the beginning of the sweep and the vertical display. However, this is rare in most digital circuit troubleshooting.

5. Adjust the controls to spread the pattern over as much of the screen as desired. Adjust the vertical position controls so that the downward swing of the waveform coincides with one of the screen lines below the centerline, as shown in Fig. 5-25.

6. Adjust the horizontal position control so that one of the upper peaks of the pulse (which presumably has a flat top) lies near the vertical centerline, as shown in Fig. 5-25.

7. Measure the peak-to-peak vertical deflection in centimeters. Multiply the distance measured by the vertical attenuator switch setting. Also include the attenuation factor (if any) of the probe. Generally, digital

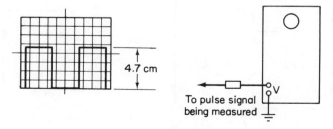

Figure 5-25 Measuring pulse amplitude.

pulse circuits are measured with a low-capacitance probe having no attenuation or where the attenuation is included as part of the vertical attenuator switch setting.

As an example, assume that a vertical deflection of 4.7 cm (Fig. 5-25) is measured, using a 1× oscilloscope probe (no attenuation), and a vertical deflection factor (step attenuator setting) of 1 V/cm. Then the peak-to-peak pulse amplitude is 4.7 × 1 × 1 × 4.7 V.

5-2-4 Measuring pulse width or duration

Pulse width or pulse duration is measured on an oscilloscope, preferably a laboratory type, where the horizontal scale is calibrated directly in relation to time. The horizontal sweep circuit of a laboratory oscilloscope is usually provided with a selector control that is direct-reading in relation to time. That is, each horizontal division on the screen has a definite relation to time at a given position of the horizontal sweep rate switch (such as 1 μs/cm). Typically, the pulses used in digital circuits are on the order of nanoseconds or microseconds, but can be as long as a few milliseconds.

The basic test connection and corresponding oscilloscope display for pulse width measurement is shown in Fig. 5-26. The basic procedure is as follows:

1. Connect the equipment as shown in Fig. 5-26. Place the oscilloscope in operation.
2. Set the vertical step attenuator to a deflection factor that will allow the expected signal to be displayed without overdriving the vertical amplifier. Set the input selector to measure ac.
3. Connect the probe to the pulse signal being measured. Switch on the oscilloscope internal recurrent sweep. Set the horizontal sweep control to the fastest sweep rate that will display a convenient number of divisions between the measurement points (Fig. 5-26). On most oscilloscopes, it is recom-

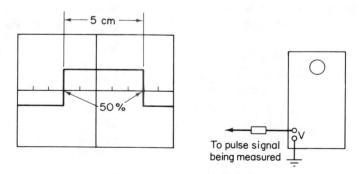

Figure 5-26   Measuring pulse width.

mended that the extreme sides of the screen not be used for width or time-duration measurements. There may be some nonlinearity at the beginning and end of the sweep.

4.   Adjust the vertical position control to move the measurement points to the horizontal centerline. Adjust the horizontal position control to align one edge (usually the left-hand edge) of the pulse with one of the vertical lines.

5.   Measure the horizontal distance between the measurement points on the left- and right-hand edges of the pulse. If width is the only concern, use the 50 percent points shown in Fig. 5-24. If rise time is to be measured, use the 10 and 90 percent points. Better resolution is obtained if the 10 and 90 percent points are used.

6.   Multiply the distance measured by the setting of the horizontal sweep control. Also include the sweep magnification factor (if any). Because of the very short pulse duration, it is often necessary to use sweep magnification.

As an example, assume that the horizontal sweep rate is 0.1 ms/cm, the sweep magnification is 100, and a horizontal distance of 5 cm is measured (between the 50 percent points). Then the pulse width is $5 \times 0.1/100 = 0.005$ ms or 5 μs.

5-2-5   Measuring pulse frequency or repetition rate

Pulse frequency or repetition rate is measured on an oscilloscope, preferably a laboratory type, where the horizontal scale is calibrated directly in relation to time. Pulse frequency is found by measuring the time duration of one complete pulse cycle (not pulse width) and then dividing the time into 1 (since frequency is the reciprocal of time).

The basic test connection and corresponding oscilloscope display for pulse-frequency measurement is shown in Fig. 5-27. The basic procedure is as follows:

1.   Connect the equipment as shown in Fig. 5-27. Place the oscilloscope in operation.

2.   Set the vertical step attenuator to a deflection factor that will allow the expected signal to be displayed without overdriving the vertical amplifier. Set the input selector to measure ac.

3.   Connect the probe to the pulse signal being measured. Switch on the oscilloscope internal recurrent sweep. Set the horizontal sweep control so that one complete pulse cycle is displayed. Avoid using the extreme sides of the screen. If necessary, display more than one complete cycle.

4.   Adjust the vertical position control to move the measurement points to the horizontal centerline. Adjust the horizontal position control to align one measurement point on the pulse with one of the vertical lines. In

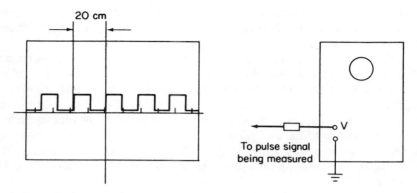

Figure 5-27   Measuring pulse frequency or repetition rate.

Fig. 5-27, the start of the left-hand edge of the pulse is used as the measurement point.

5.   Measure the horizontal distance between the measurement points. Multiply the distance measured by the setting of the horizontal sweep control. Also include the sweep magnification factor (if any). Divide the time measured into 1 to find frequency.

As an example, assume that the horizontal sweep rate is 0.1 ms/cm, the sweep magnification is 100, and a horizontal distance of 20 cm is measured between the beginning and end of a complete cycle. Then the frequency is 20 × 0.1/100 = 0.020 ms or 20 μs; 1/20 μs = 50 kHz.

5-2-6   Measuring pulse times with external timing pulses

It is possible to use an external *time-mark generator* to calibrate the horizontal scale of an oscilloscope for pulse-time measurement. Time-mark generators are used for many reasons, but their main advantage for digital applications are greater accuracy and resolution. For example, a typical crystal-controlled, time-mark generator will produce timing signals at intervals of 10 and 1 μs, as well as 100 ns.

A time-mark generator produces a pulse-type timing wave, which is a series of sharp spikes placed at precise time intervals. These pulses are applied to the oscilloscope vertical input and appear as a wavetrain similar to that shown in Fig. 5-28. The oscilloscope horizontal gain and positioning controls are adjusted to align the timing spikes with screen lines, until the screen divisions equal the timing pulses. The accuracy of the oscilloscope timing circuits is then of no concern, since the horizontal channel is calibrated against the external time-mark generator.

The timing pulses can be removed, and the pulse signal to be measured

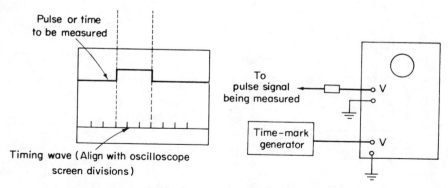

Figure 5-28 Measuring pulse time with external timing pulses.

can be applied to the vertical input, provided that the horizontal gain and positioning controls are not touched.

Duration or time is read from the calibrated screen divisions in the normal manner. If the oscilloscope has a dual-trace feature, the time-mark generator can be connected to one input; the other input receives the pulse signal to be measured. The two traces can be superimposed or aligned, whichever is convenient.

5-2-7 Measuring pulse delay

The time interval or delay between two pulses (say an input pulse and an output pulse) introduced by a delay line, FF, or other digital circuit can be measured most conveniently on an oscilloscope with a dual trace. It is possible to measure delay on a single-trace oscilloscope, but it is quite difficult. If the delay is exceptionally short, the screen divisions can be calibrated with an external time-mark generator. If the oscilloscope is a dual-trace instrument, the screen divisions must be calibrated against the time-mark generator as described in Sec. 5-2-6. Then the time-mark generator can be removed, and the input and output pulses can be applied to the two vertical channels.

The basic test connection and corresponding oscilloscope display for pulse-delay measurement is shown in Fig. 5-29. The basic procedure is as follows:

1. Connect the equipment as shown in Fig. 5-29. Place the oscilloscope in operation. Switch on the internal recurrent sweep.

2. Set the sweep frequency and sync controls for a single stationary input pulse and output pulse as shown in Fig. 5-29. Set the horizontal and vertical gain controls for the desired pulse pattern, width, and height. Avoid the extreme limits of the screen.

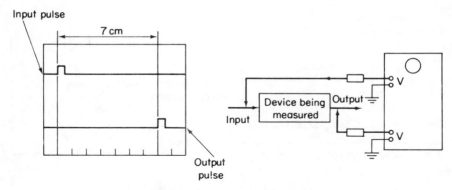

Figure 5-29   Measuring pulse delay.

3.   Count the screen divisions between the *same point* on both the input and output pulses. This is the delay interval.

As an example, assume that the horizontal sweep rate is 100 ns/cm, there are seven screen divisions between the input and output pulses, and each horizontal screen division is 1 cm. Then the delay is $7 \times 100 = 700$ ns.

5-2-8   Testing digital circuits

The following paragraphs describe procedures for test or checkout of basic digital circuits (gates, delays, FFs, etc.). Such circuits can be tested on a single-pulse basis or with pulse trains. Both test methods are described.

An oscilloscope is the best tool for pulse-train testing. In its simplest form, pulse-train testing consists of measuring the pulses at the input and output of a digital circuit. The pulses can be those normally present in the circuit, or they can be introduced from an external pulse generator. If the input pulses are normal but the output pulses are absent or abnormal, the fault is most likely to be in the circuit being tested. An exception to this is where the following circuit is defective (say a short circuit) and makes the output of the circuit under test *appear* to be absent or abnormal. Either way, the trouble is localized to a specific point in the overall circuit.

If the input pulses to the circuit under test are absent or abnormal, the trouble is most likely to be in circuits ahead of the point being tested. Thus, when input pulses are found to be abnormal at a certain circuit, this serves as a good starting point for troubleshooting.

The logic probe (Sec. 5-1-5) can also be used for pulse-train analysis. However, the logic probe indicates only that a pulse train exists at a given point. The logic probe cannot tell the pulse width, frequency, amplitude, or whether the pulse is in step with other pulses. Only an oscilloscope will show this information.

The logic probe and logic clip are ideal tools for single-pulse analysis. Such testing can take several forms. The clock frequency can be slowed down to some very low rate (typically 1 Hz or slower), so that the individual input and output pulses can be monitored individually (on the probe, clip, or oscilloscope). The clock rate is slowed by substituting an external pulse generator for the normal (internal) clock signal. If this is not practical, it is possible to inject pulses from the generator directly at the inputs of a circuit, or pulses can be introduced with a logic pulser (Sec. 5-1-6).

Keep in mind that digital troubleshooting usually starts with a test of several circuits simultaneously. There may be several hundred (or thousand) circuits in a digital device. It could take months to check each circuit on an individual basis. However, there are tests that check whole groups of circuits simultaneously. The self-check or operational-check procedures found in digital equipment service literature are based on this technique. If the digital device is programmed, it is possible to use test programs (also known as *diagnostic routines*) to check large groups of circuits, and large sections of the program, for proper operation. Such diagnostic routines are discussed in Sec. 5-5.

A classic example of testing several digital circuits simultaneously is when four FFs are connected as a decade counter. If there is an output pulse train at the fourth FF, and this pulse train shows one pulse for every 10 input pulses at the first FF, the entire counter can be considered as good.

Also keep in mind that most present-day digital circuits are combined in IC packages. Even if it is possible to test the individual circuits, they cannot be replaced. The complete IC function must be checked as a package by monitoring inputs and outputs at the IC terminals.

Marginal test. Before going into detailed test procedures, the marginal test technique should be considered. The power supplies for many digital circuits are adjustable, usually over a ± 10 percent range. When digital equipment failure is intermittent or the failure is not consistent, it is sometimes helpful to test the circuits with the power-supply output set to extremes (both high and low). This will show marginal failures in some cases. The technique is not recommended in all digital service literature. However, it is usually safe to apply the technique to any equipment on a temporary basis. Of course, the equipment service literature should be consulted to make sure.

The following procedures are based on the assumption that the corresponding digital circuit can be tested on an individual basis.

5-2-9 Testing AND gate circuits

Figure 5-30 shows the basic connections for testing an AND circuit. Ideally, both inputs and the output should be monitored simultaneously (with a multitrace oscilloscope), since an AND circuit produces an output

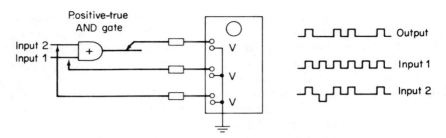

Figure 5-30   Testing AND gate circuits.

only when two inputs are present. If pulse trains are monitored on an oscilloscope, check that an output pulse is produced each time there are two input pulses of appropriate amplitude and polarity. If this is not the case, check carefully that both input pulses arrive at the same time (not delayed from each other).

If the oscilloscope has a dual-trace feature, first check for a series of simultaneous pulses at both inputs. Note that in Fig. 5-30 the input pulse trains are not identical, which is often the case. Input 2 has fewer pulses for a given time interval than input 1. Also, input 2 has one negative pulse simultaneously with one positive pulse at input 1. This should not produce an output pulse. Often, one input will be a long pulse (long in relation to the pulses at the other input). In other cases, one input will be a fixed d-c voltage applied by operation of a switch.

If the AND circuit is tested with pulse trains and a logic probe or clip, simply check for a pulse train at both inputs and at the output. If the circuit is checked on a single-pulse basis, inject simultaneous pulses at both inputs and monitor the output with probe or clip.

The complete truth table of the AND gate can be checked using the basic connections of Fig. 5-30. For example, an output pulse train (or single pulse) should not appear if only one input is present.

5-2-10   Testing OF gate circuits

Figure 5-31 shows the basic connections for testing an OR circuit. Ideally, all inputs and outputs should be monitored simultaneously. An OR gate produces an output when any inputs are present. If pulse trains are monitored on an oscilloscope, check that an output pulse is produced for each input pulse of appropriate amplitude and polarity.

If the oscilloscope has a dual-trace feature, monitor the output with one trace. Then monitor each of the inputs in turn with the other trace. Note that in Fig. 5-31 the input pulse trains are not identical. That is, the input pulses do not necessarily coincide. However, there is an output pulse when either input has a pulse and when both inputs have a pulse. This last condi-

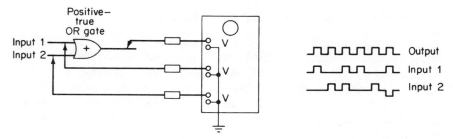

Figure 5-31 Testing OR gate circuits.

tion marks the difference between an OR gate and an EXCLUSIVE OR gate, discussed in Sec. 5-2-11.

If the OR circuit is tested with pulse trains and a logic probe or clip, check that there is a pulse train at the output whenever there is a pulse at any input. If the circuit is checked on a single-pulse basis, monitor the output and inject a pulse at each input, in turn. The complete truth table of the OR gate can be checked using the basic connections of Fig. 5-31.

5-2-11 Testing EXCLUSIVE OR gate circuits

Figure 5-32 shows the basic connections for testing an EXCLUSIVE OR circuit. Ideally, all inputs and the output should be monitored simultaneously. An EXCLUSIVE OR gate produces an output when any one input is present, but not when both inputs are present. If pulse trains are monitored on an oscilloscope, check that an output pulse is produced for each input pulse that does not coincide with another input pulse.

If the EXCLUSIVE OR circuit is checked with a logic probe or clip, do so on a single-pulse basis. The logic probe or clip can indicate the presence of pulse trains, but not coincidence of pulses in pulse trains. Monitor the output and inject a pulse at each input, in turn. An output pulse should be produced for each input pulse. Then apply simultaneous pulses to both inputs. There should be no output. The complete truth table of the EXCLUSIVE OR gate can be checked using the basic connections of Fig. 5-32.

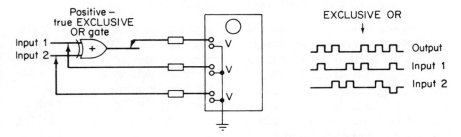

Figure 5-32 Testing EXCLUSIVE OR gate circuits.

5-2-12   Testing EXCLUSIVE NOR gate circuits

Figure 5-33 shows the basic connections for testing an EXCLUSIVE NOR circuit. Ideally, all inputs and the output should be monitored simultaneously. An EXCLUSIVE NOR gate produces an output when both inputs are present, or both are absent, but not when the inputs are mixed (one present, one absent). If pulse trains are monitored on an oscilloscope, check that an output pulse is produced only when both inputs are present or both are absent.

If the EXCLUSIVE NOR circuit is checked with a logic probe or clip, do so on a single-pulse basis. The logic probe or clip can indicate the present of pulse trains, but not coincidence of pulse in pulse trains. Monitor the output with no pulses at either input. The output should show a steady high condition. Then monitor the output and inject a pulse at each input, in turn. There should be no output pulse produced for either input pulse. Finally, apply simultaneous pulses to both inputs. There should be an output. The complete truth table of the EXCLUSIVE NOR gate can be checked using the basic connections of Fig. 5-33.

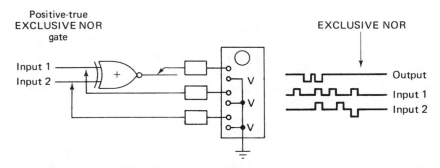

Figure 5-33   Testing EXCLUSIVE NOR gate circuits.

5-2-13   Testing NAND and NOR gate circuits

Figure 5-34 and 5-35 show the basic connections for testing NAND and NOR circuits, respectively. Note that the connections are the same for corresponding AND and OR circuits, as are the test procedures. However, the output pulse is inverted. That is, a NAND circuit produces an output under the same logic conditions as an AND circuit, but the output is inverted. For example, if the logic is positive true and a positive pulse is present at both inputs of a NAND gate simultaneously, there will be an output pulse, and the pulse will be negative (or false).

342

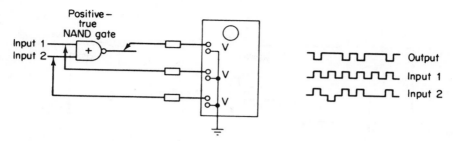

Figure 5-34 Testing NAND gate circuits.

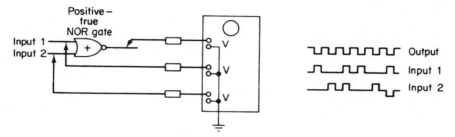

Figure 5-35 Testing NOR gate circuits.

5-2-14 Testing amplifiers, inverters, and phase splitters

Figure 5-36 shows the basic connections for testing amplifiers, inverters, and phase splitters used in digital circuits. Note that the same connections are used for all three circuits. That is, each circuit is tested by monitoring the input and output with an oscilloscope or logic probe. However, the relationship of output to input is different for each of the three circuits.

With the amplifier, the output pulse should be greater than the input pulse amplitude. With the inverter, the output pulse may or may not be greater in amplitude than the input pulse, but the polarity will be reversed. In the case of the phase splitter, there will be two output pulses of opposite polarity to each other for each input pulse.

Polarity is not difficult to determine if the pulses are monitored on an oscilloscope. The ability of a logic probe to show polarity depends on the system used. A typical logic probe will respond to systems where the logic levels are between 0 and 5 V. That is, a true or high level is 5 V, while a false or low level is 0 V. The logic probe will blink on for each pulse, if the pulse goes from some level below about 1.4 V to a level above 1.4 V, and will blink off when the pulse drops from a high level to some voltage below about 1.4 V. If the probe is used to check an inverter where the input goes from 0 to 5 V

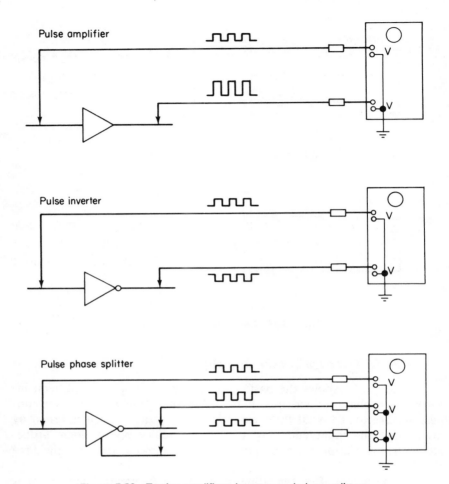

Figure 5-36   Testing amplifiers, inverters and phase splitters.

and the output goes from 5 to 0 V, the input pulse train will produce a series of "on blinks" while the output pulse train produces "off blinks." It is difficult, if not impossible, to tell the difference between on and off "blinks" if the pulse trains are fast. Also, if an inverter produces a negative output that goes from 0 to -5 V (when a $+5$ V input is applied) a typical logic probe will not respond properly.

5-2-15   Testing FF circuits

Figure 5-37 shows the basic connections for testing an FF. As discussed in Chapter 2, there are many types of FFs (RS, JK, toggle, latching, delay, etc.). Each FF responds differently to a given set of pulse conditions. Thus, each type of FF must be tested in a slightly different way. However, all FFs have two states (even though there may be only one output, and it may not be possible to monitor both states directly). Some FFs require only one input to change states. Other FFs require two simultaneous pulses to change states (such as a clock pulse and a reset or set pulse).

The main concern in testing any FF is that the states change when the appropriate input pulse (or pulses) are applied. Thus, at least one input and one state (output) should be monitored simultaneously. For example, an RS FF (without clock) can be tested by monitoring the pulse train at the set (S) input and set (D) output, as shown in Fig. 5-37. If there is a pulse train at the S input and at the D output, it is reasonable to assume that the FF is operating. If the D output remains in one state, either the FF is defective and not being reset or there are no reset pulses. The next step is to monitor the R (reset) input and D output. If both the S and R input pulse trains are present but the FF does not change states, the FF is defective.

There are exceptions of course. For example, if the reset and set pulses arrive simultaneously due to an unwanted delay in other circuits, the FF may remain in one state.

If an FF requires a clock pulse, first check that both the clock pulse and an input pulse arrive simultaneously. Then check that the FF changes states each time there is such a pulse coincidence. For example, assume that pulse trains are applied to the JK FF of Fig. 5-37, and that the pulses are monitored on a multitrace oscilloscope. Also assume that a (J) input pulse occurs for

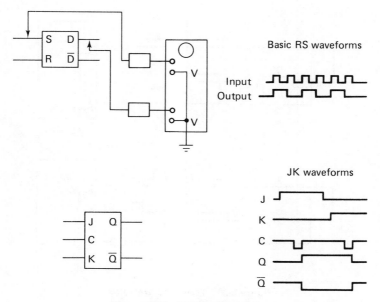

Figure 5-37 Testing FF circuits.

every other clock (C) pulse. Monitor the C pulse and the Q output pulse, and note that the Q output occurs once for every two C input pulses. Or, monitor the J input and Q output, noting that the FF changes states on a pulse-for-pulse basis.

To test an FF with pulse trains and a logic probe or clip, simply check that a pulse train exists at each output, with a pulse train at the inputs. It can then be assumed that the FF is operating properly. To remove all doubt, use the logic probe to test the FF on a single-pulse basis.

5-2-16 Testing MVs

Figure 5-38 shows the basic connections for testing an MV. There are three basic types of MVs: free-running, one-shot (monostable), and Schmitt trigger (bistable). In the case of the free-running MV, Only the output need be monitored since the circuit is self-generating. Both the one-shot (OS) and Schmitt trigger (ST) require that the input and output be monitored.

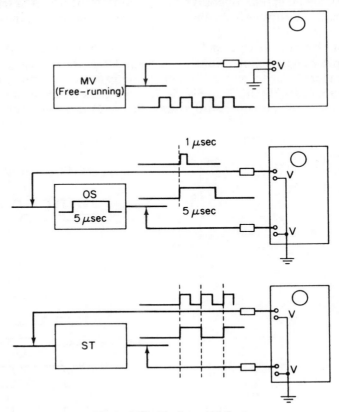

Figure 5-38 Testing multivibrators.

If a logic probe or clip is used to monitor pulse trains from an MV, the presence of a pulse train at the output indicates that the MV is *probably* working. The OS and ST MVs can be tested on a single-pulse basis as well by injecting pulses at the input. Keep in mind that the ST requires two input pulses for a complete cycle. That is, two input pulses are required to change the output states and then return the states back to the original condition. More simply, two input pulses are required for one complete output pulse. The OS MV produces one complete output pulse for each input pulse, even though the output pulse may be different in duration from that of the input pulse.

If MV pulse trains are monitored on an oscilloscope, the frequency, pulse duration, and pulse amplitude can be measured as discussed in Secs. 5-2-3 through 5-2-5. While these factors may not be critical for all digital circuits, quite often one or more of the factors is important. For example, the logic symbol for an OS MV (when properly drawn) includes the output pulse duration or width.

5-2-17   Testing delay elements

Figure 5-39 shows the basic connections for testing delay lines or delay elements used in some digital circuits. The test procedure is the same as for measurement of delay between pulses described in Sec. 5-2-7. Keep in mind that the delay shown on the logic symbol usually refers to the delay between leading edges of the input and output pulses. This can be assumed if the symbol is not further identified. Some manufacturers identify their delay symbols as to leading or trailing edge, particularly if the symbols are mixed on a single logic diagram.

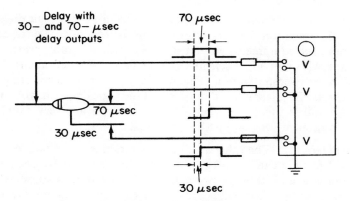

Figure 5-39   Testing delays.

5-3 DIGITAL EQUIPMENT SERVICE LITERATURE

Unlike some other electronic equipment, it is almost impossible to troubleshoot digital circuits without adequate service literature. There are two basic reasons for this.

Although all digital devices use the same basic circuits (gates, FFs, etc.), the way in which these circuits are arranged to perform an overall function is unique to each type of digital equipment. Thus, it is absolutely essential that you know exactly what the digital equipment will do from an operational standpoint. Further, you must know what pulse will appear at what point and under what condition.

Another reason for good service literature is that, in servicing digital equipment, you are dealing with failure rather than poor performance. For example, a television set may have a weak picture or some distortion in the audio. This may pass unnoticed or be ignored during normal operation. On the other hand, a digital counter will either read out a correct count, fail to read out a correct count, or provide no readout. Anything less than the correct count is not acceptable. You must know what is correct in every case. Knowing the equipment can come only after a thorough study of the service literature. In the case of very complex digital devices (particularly large computers, graphic terminals, etc.) you must also receive factory training before you can hope to do a good troubleshooting job.

5-3-1 Important information found in digital service literature

All of the information found in service literature is of value, but the most important pieces of information for troubleshooting are the logic diagram, the theory of operation, and the self-check procedure. Also, timing diagrams can be of considerable help.

Logic diagram and theory of operation. The value of the logic diagram and theory of operation for troubleshooting is obvious. To trace pulses throughout the equipment, you must have a logic diagram, preferably supplemented with individual schematics that show the internal circuits of the blocks on the overall diagram. The individual schematics are omitted when the blocks are IC packages. To understand the logic diagram, you must have a corresponding theory of operation (unless you are very familiar with similar equipment, or you just happen to be a genius).

Timing diagrams. The value of a timing diagram is sometimes less obvious. However, trouble can often be pinpointed immediately by comparison of circuit pulses against those shown in the timing diagram. Also, unusual problems can show up when pulses are compared with the timing diagram. For example, assume that a counter receives input (or count) pulses and reset pulses, both of which originate from a common clock or master oscillator. The reset pulses are supposed to arrive at the counter at some specific time interval, thus allowing a given maximum number of clock pulses (say, 100) to be read out. Further, assume that a defect in the circuit has caused the reset pulsed to be delayed slightly in relation to the clock pulse so that the reset pulse coincides with the first count. The counter will never be able to reach a second count, and the readout will always remain at zero, even though all of the circuits appear to be good.

Self-check.   The value of a self-check procedure depends on the complexity of the equipment. On simple equipment, an ingenious technician can sometimes devise a check procedure that will eliminate groups of digital circuits from suspicion or pinpoint the problem to a group of circuits. For example, assume that a digital counter is to be checked. The counter gate can be set to open for 1-second intervals, with a 1-kHz pulse signal introduced at the counter input. If the counter is operating properly, the readout will be 1000. (From a practical standpoint, if the readout is 0999 or 1001, the counter may still be functioning properly. Most digital counters have a ± 1 count ambiguity.) The gate time is then shortened or lengthened, and the input pulse frequency is changed (increased or decreased), to check all of the digits in the readout or each digit one at a time, whichever is convenient.

On complex equipment such as a computer, the self-check must be thought out carefully. Unless you are completely familiar with every logic circuit, you will do much better to follow the service literature. Usually, a computer or logic analyzer self-check involves running the instrument through a diagnostic routine and then noting the readout. The readouts are compared line for line, againt the routine, until an abnormal readout is found. The service literature often contains some means of relating the routine and/or readout to a specific circuit or group of circuits.

5-3-2   Examples of service literature

The following paragraphs show typical examples of information found in digital equipment service literature. This information is extracted from the manual for the Heathkit H10 Paper-Tape Reader/Punch. This information is summarized; full information appears in the manual. However, the information presented here is sufficient to familiarize the reader with what is available in well-prepared service literature, and how the data can be related to troubleshooting.

Mechanical details.   Figure 5-40 shows basic mechanical details of the reader/punch, which both reads and punches standard eight-channel, 1-inch (10-inch-maximum-diameter roll or fanfold) paper tape. The reader and punch may be operated simultaneously and yet controlled independently.

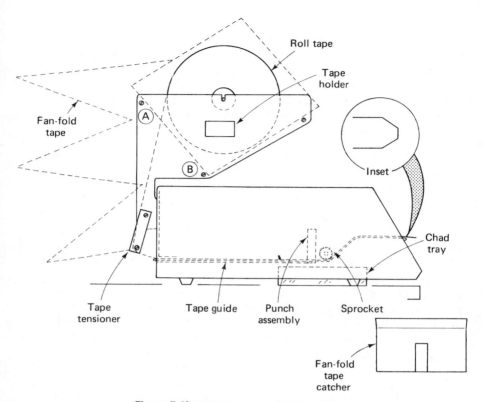

Figure 5-40 Mechanical details of tape punch.

The H10 has a standard TTL parallel interface, a pushbutton feed switch to generate leader tape, uses a photoelectric tape reader (Sec. 3-6-1), and has a stepper motor for the tape reader transport.

Warning: Some types of paper tape contain toxic chemicals. Therefore, be careful not to use the tape such that it can contact a person's mouth or be inhaled. Be especially careful of the chad, or holes punched from the tape, since these tiny bits can be picked up on the fingers. This warning applies to all types of paper-tape reader/punches used with digital electronics.

Photoelectric tape reader circuits. There are eight tape reader circuits, one for each of the eight channels on the paper tape. All eight reader circuits operate in the same manner. Therefore, only one circuit, shown in Fig. 5-41, is discussed.

As a hole in the paper tape is positioned between the lamp LP1 and the photo-Darlington transistor Q101, the light causes Q101 to turn on for the

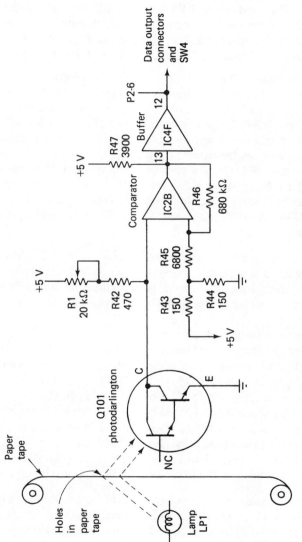

Figure 5-41 Typical tape reader circuit.

time that the tape hole is between the lamp and Q101. This causes the output of comparator IC2B to go high. In turn, this causes buffer IC4F to go high. The signal is then applied to the data output connectors and switch SW4.

Resistor R42 and control R1 set the sensitivity of photo-Darlington transistor Q101. Resistors R43 and R44 set the reference trigger level of comparator IC2B. R47 is an output pull-up resistor. Resistors R45 through R47 provide positive feedback for IC2B to eliminate false triggering.

Tape motor timing. Timing for the stepper motor drive circuitry is generated by one-shot multivibrator IC14, shown in Fig. 5-42. When READ switch SW7 is pushed ON, pin 12 of IC13D goes low. This causes pin 11 to go high and trigger IC14, if the reader-start line (P3-10) is low. Pin 6 is then driven low, and this low is applied to IC15 (pin 12) of the motor drive circuitry and pin 9 of reader-ready gate IC13C.

At this time, IC13C pin 8 goes high and drives IC7D, which indicates that the reader is busy. After IC14 times out, $\overline{Q}$ (pin 6) goes high again, and waits for another 1-to-0 transition from the reader start line (IC14, pin 1). When READ switch SW7 is ON, a 1-to-0 on IC14 pin 1 will advance the tape one character.

Motor drive. IC15 divides the timing signal at pin 12 (either from the reader timing circuits or punch timing) by 2, and produces two output signals (at pins 8 and 6) that are 180° out of phase with each other. The remaining circuitry produces a four-phase signal to drive the stepper motor as shown in Fig. 5-42.

When power is applied, assume that IC15 $\overline{Q}$, IC16B Q, and IC16A $\overline{Q}$ are high. Then, because the D inputs of IC16 Q are tied to IC16A, IC16A $\overline{Q}$ is driven high, and the timing sequence shown in Fig. 5-43 begins.

With the first pulse to IC15T (pin 12), IC15 Q goes high and $\overline{Q}$ goes low. Because of the high on the D inputs of IC16, IC16B goes high and $\overline{Q}$ goes low. IC16A remains unaffected by the transition. The next time, IC15 changes state, IC16B remains unaffected, and IC16A changes. As the input pulses continue, a four-phase signal is produced that is applied through buffers A, B, C, and D of IC4 to drive stepper motor transistors Q23–Q26. Diodes D21–D24 suppress the inductive kickback from the motor windings.

Tape punch. Because the tape punch circuits are so similar, only channel 1 (shown in Fig. 5-44) is described. Tape punch signals (from either the tape reader circuits or the data input connector) are coupled through switch SW4 to AND gate IC6A. When the signal at pin 2 is high and the punch timing signal at pin 1 is high, pin 3 goes high and turns on Darlington transistor pair Q17 and Q18. This allows current to flow through solenoid L1, and a hole is punched in the paper tape. Diodes D10 and D19 suppress the inductive kickback of L1. Diode D10 also causes L1 to release energy quicker than nor-

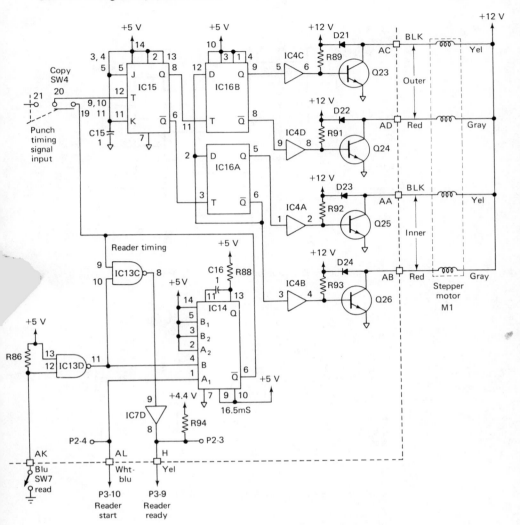

Figure 5-42   Tape motor drive and reader timing circuits.

mal by waiting until the inductive kickback reaches approximately 20 V above the supply before zener action occurs and suppression begins.

Power supplies.   The power-supply circuits are shown in Fig. 5-45.

+25-, +12-, and +9-V supplies: a-c voltage from the secondary winding of transformer T1 is rectified by diodes D2 and D7, and filtered by capacitor C3 to produce +25 V. The +12- and +9-V supplies operate in a similar manner.

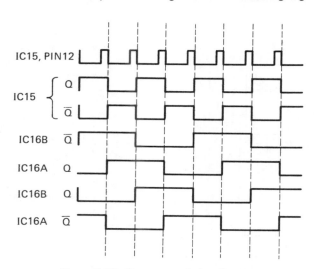

Figure 5-43 Tape motor timing diagram.

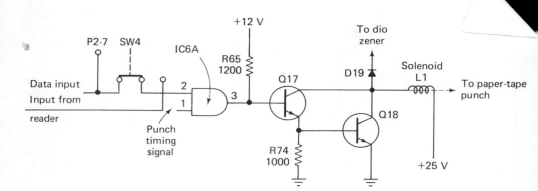

Figure 5-44 Typical tape punch circuit.

+5- and +4.4-V supplies: IC17 is a voltage regulator that changes +9 V to +5 V. Capacitors C7 and C8 provide noise suppression. Diode D25 keeps externally connected equipment from supplying power to the +5-V supply and turning on some of the output buffers when the reader/punch is off.

Lamp supply: the constant zener voltage of diode D1 is applied to the base of transistor Q1 and holds the Q1 emitter constant. In turn, this holds the base and emitter of transistor Q2 constant, and the lamp LP1 receives a constant voltage independent of variations in the +9-V supply.

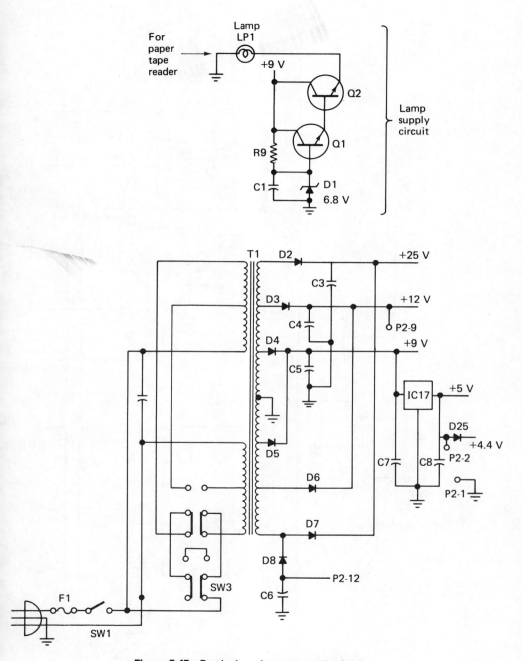

Figure 5-45 Reader/punch power supply circuits.

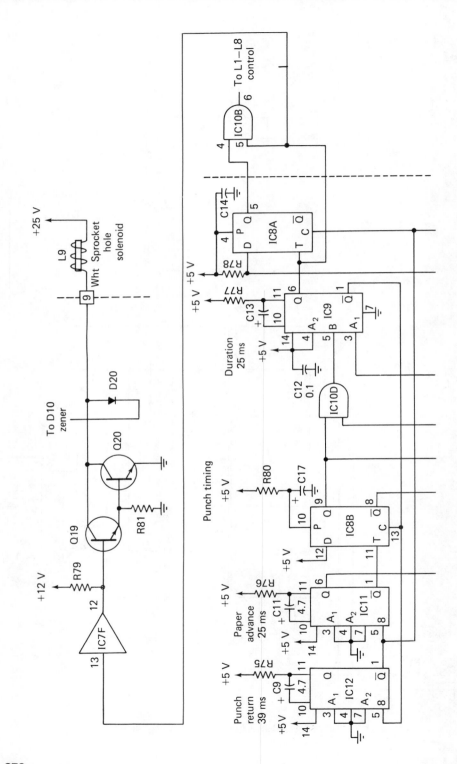

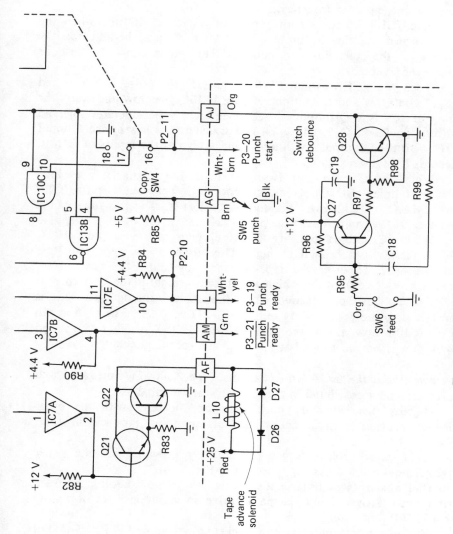

Figure 5-46 Punch timing circuits.

Punch timing. As shown in Fig. 5-46, three one-shot multivibrators control the punch timing. IC9 controls the direction of the punching operation, IC12 provides the time for the punches to return to their normal positions, and IC11 times the paper-advance solenoid. The multivibrators are interconnected so they will perform their timing functions in the proper sequence.

Gate IC10B is for circuit protection. Without IC10B, when the unit is turned on, the Q output of IC8A could be high, and possibly turn on all the solenoids continuously. This could blow the fuse or damage the solenoids.

When the FEED switch SW6 is pushed, the pulse is debounced by Q27 and Q28, and the output of gate IC10C goes low and starts the timing of IC9 (the Q output of IC9 goes high). This operates solenoid L9 (through IC7F, Q19, and Q20) and transfers the low at the D input of IC8A to the Q output. With a low and a high at the inputs of IC10B, solenoids L1–L8 remain off.

When the Q output of IC9 goes high, the $\overline{Q}$ output goes low. After IC9 times out, the $\overline{Q}$ output goes high again and starts IC12, driving the $\overline{Q}$ of IC12 low. This does not affect IC11 at this time, but it does clear IC8A and turns off the punch drivers.

When IC12 times out, its $\overline{Q}$ output goes high and starts IC11, driving the Q output of IC11 high. This drives IC7A and transistors Q21 and Q22, to drive the tape advance solenoid L10. When IC11 times out, its $\overline{Q}$ output goes high and forces a high at the Q output of IC8B, which has been cleared when IC9 started timing out. This is transferred to the punch ready output (P3-21) by IC7E. The high at pin 13 of IC10D now allows the PUNCH and FEED switches (SW5 and SW6), or a *punch start* signal at P3-19, to control the punching.

Troubleshooting the reader/punch circuits. The following notes are based on recommendations found in the reader/punch manual.

If only one solenoid punch does not operate, with the other solenoid punches normal, check the corresponding tape circuit (Fig. 5-44).

If all punch solenoids are continuously activated but the paper drive solenoid is not operating, check IC9 (Fig. 5-46).

If all punch solenoids (except sprocket hole solenoid L9, Fig. 5-46) are continuously activated but the paper drive solenoid operates normally, check IC10B (Fig. 5-46).

If all punch solenoids (except sprocket hole solenoid L9, Fig. 5-46) fail to energize but the paper drive solenoid operates normally, check IC8A.

If all punches are inoperative, check the +25-V supply (Fig. 5-45). Then check IC8B, IC9, IC10C, IC10D, and IC13B.

If the punches and motor M1 (Fig. 5-42) are inoperative, check the +12-V supply (Fig. 5-45).

If all solenoids are energized, the tape reader is inoperative, the reader lamp LP1 lights, but the fuse F1 (FIg. 5-45) blows in a short time, check IC17 and the +5-V supply.

If the punch solenoids energize for only one character after power is appled, check IC11 and IC12 (Fig. 5-46).

If the sprocket teeth tear the punched tape, check for binding in the solenoid linkage, for tape that is not properly threaded, for a drive sprocket that is incorrectly aligned, and for tape tensioner binding. Also check IC12 (Fig. 5-46).

If the chad is not completely removed from the tape, check for punch solenoids incorrectly adjusted, unsuitable tape, and worn die block assembly. Also check for low line voltage and a defective IC9 (Fig. 5-46).

If the tape does not advance or is erratic but the tape advance solenoid L10 actuates normally, check that the detent or solenoid is properly adjusted. Also check if one or more punches are not returning to their released positions.

If the tape-advance solenoid L10 (Fig. 5-46) does not operate but the punches operate normally, check L10, IC7A, Q21, Q22, and IC11. Also check for binding in the ratchet mechanism.

If the tape reader lamp LP1 does not light but the tape advances normally, check for a burned-out lamp LP1 or for defective transistors Q1 and Q2 (Fig. 5-45).

If only one data channel does not change state at a tape hole, with other data channels normal, check the corresponding tape reader circuit (Fig. 5-41). Also check for dirt in the light passage to Q101 and for proper adjustment of R1.

If the stepper motor (Fig. 5-42) is inoperative or erratic in both READ and COPY modes but the punch operates normally, check M1, Q23-Q26, IC14, IC15, and LC16. If the stepper motor runs backward, check that the red and black leads are not interchanged.

If the stepper motor is inoperative in the READ mode but operates normally in the COPY mode, check IC13D and IC14 (Fig. 5-42).

If all outputs remain low with no tape in the reader and adjustments R1–R8 (Fig. 5-41) have no effect, check R43. Also check the position of the lamp.

If all outputs remain high with blank tape in the reader and adjustments of R1–R8 have no effect, check R44 (Fig. 5-41). It is also possible that the tape is too transparent.

If there are consistent reading errors, check for marginal adjustment of R1–R8 (Fig. 5-41), marginal adjustment of tape to photocell housing, incorrect tape hole spacing, tape holding spring too high, excessive drag on unread tape supply, damaged tape, dirt in photocell housing, or tape too transparent. It is also possible that IC14 (Fig. 5-42) in the reader timing circuit is defective, but a mechanical defect is more likely. (It has been the author's experience that technicians have more difficulty in locating and correcting mechanical defects, such as occur in tape reader/punches, floppy disks, and cassettes, than defects in the electronic circuits.)

5-4   DIGITAL IC TEST AND TROUBLESHOOTING

The circuits of a typical digital device are mostly IC rather than discrete. As in the case of a linear IC, you do not have access to the internal circuits of a digital IC. Thus, you must determine if the IC is performing its function, or failing, without actually getting into the circuits. (From a practical standpoint, it is very helpful if you can make this decision before you pull the IC from the printed-circuit board.) Unfortunately, the traditional tools for discrete digital troubleshooting often provide little or no help in making these decisions in the case of digital ICs. Let us consider two classic examples of digital troubleshooting techniques, the comparison of timing functions, and point-to-point signal pulse tracing.

5-4-1   Timing in digital IC troubleshooting

An often-used method for troubleshooting of discrete digital circuits is to introduce a pulse train at the circuit input while monitoring various points throughout the circuit with an oscilloscope. If any of the pulses are absent or abnormal (such as very low amplitude pulses or pulses that occur at the wrong time), this means a failure or degrading of the components between the pulse source (pulse generator) and monitor (oscilloscope). For example, a degraded diode or transistor can introduce an abnormally long delay and cause an output pulse to occur too late to open a buffer. A leaking capacitor can appear as "pulse jitter" on the oscilloscope. However, when a digital IC fails, there is total failure. Digital IC timing parameters rarely degrade or become marginal. Thus, critical timing measurements are usually not too important in troubleshooting.

This does not mean that timing measurements are to be omitted from digital troubleshooting. Timing measurements are often critical during the design and development stage of digital systems. Generally, it is a case where some IC (or group of ICs) does not operate fast enough to perform each program step, as initially programmed. Such problems are solved by debugging (changing the program sequence) rather than by troubleshooting. As a general rule, once a program has been debugged and the digital device performs properly, timing will remain good (or will fail completely).

There is an exception to this general rule. Digital ICs tend to speed up when power-supply voltages are high and slow down with low power-supply voltages. Thus, if program timing is on the borderline and there is a drastic change in power-supply voltage (particularly if the voltage is low), system timing can result in improper timing.

5-4-2   Point-to-point pulse tracing

Point-to-point pulse tracing can be used in digital troubleshooting. Pulses are introduced on a line or at some point in the circuit (with a pulse generator or logic pulser) and then monitored at another point on the line (with an oscilloscope, logic probe, or clip). Unfortunately, there are certain cases where this does not work too well with digital ICs.

Consider the circuit of Fig. 5-47, which is part of the output buffer (connected to one line of the data bus) in a microcomputer RAM. (The circuit is sometimes referred to as a "transistor totem pole.") In either the high or low (1 or 0) state, the circuit is a low impedance. In the low state, the circuit output is a saturated transistor to ground, and appears as about 5 or 10 ohms to ground. This presents a problem in pulse tracing. A signal source used to inject a pulse at a point that is driven by this output (such as trying to inject a pulse on that line of the data bus) must have sufficient power to override the low impedance output state.

A typical pulse generator (even a logic pulser) may not have this capability. Instead, it is necessary to either cut printed-circuit traces (or lines) or pull IC leads in order to inject the pulses. Both of these practices are time-consuming and lead to unreliable repairs. However, there are basic digital IC

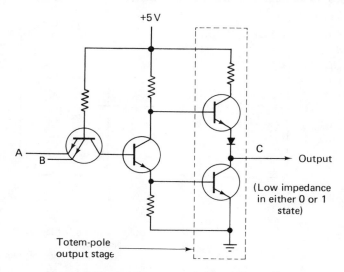

Figure 5-47 Output buffer with totem-pole output.

troubleshooting techniques, using logic probes and pulsers, as well as current tracers, that will resolve such problems. These techniques are discussed in the following paragraphs. But first let us consider the how and why of digital IC failure.

5-4-3   How digital ICs fail

It is essential that you understand the types of failure that occur in digital ICs if you are going to troubleshoot a digital device efficiently. There are two classes of digital IC failure: (1) those caused by internal failure, and (2) those caused by a failure in the digital circuit external to the IC.

Four types of failure can occur internally to a digital IC: (1) an open bond on either an input or output, (2) a short between an input or output and V_{CC} or ground, (3) a short between two pins (neither of which is V_{CC} or ground, and (4) a failure in the internal circuitry (often called the steering circuitry) of the IC.

Four types of failure can occur in the circuit external to the IC. These are (1) a short between a point (or *node,* as it may be called in digital literature) and V_{CC} or ground, (2) a short between two nodes (neither of which are V_{CC} or ground), (3) an open signal path, and (4) a failure of an external component.

Effects of an open output bond.   When there is an open output bond, as shown in Fig. 5-48, the inputs driven by that output are left to float. In a typical TTL IC, a floating input rises to approximately 1.4 to 1.5 V and usually has the same effect on circuit operation as a high logic level. Thus, an

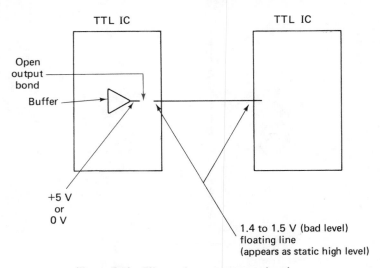

Figure 5-48 Effects of an open output bond.

open output bond will cause all inputs driven by that output to float to a bad level, since 1.5 V is less than the high threshold level of 2 V and greater than the low threshold level of 0.4 V. In TTL ICs, a floating input is usually interpreted as a high level. Thus, the effect will be that these inputs will respond to this bad level as though it were a static high level.

Effects of an open input bond. When there is an open input bond, as shown in Fig. 5-49, we find that the open circuit blocks the signal driving the input from entering the circuit (a microprocessor data bus buffer, in this case). The input line is thus allowed to float, and will respond as though it were a static high signal. It is important to realize that, since the open occurs on the input *inside* the IC, the digital signal driving this input will be unaffected by the open. That is, the signal is blocked to the microprocessor buffer inside the IC, but there will be no effect on other ICs. Of course, the microprocessor will respond as if there is a static 1 on that line of the data bus and will thus not respond properly to those commands or words where a 0 is to appear on that line of the bus.

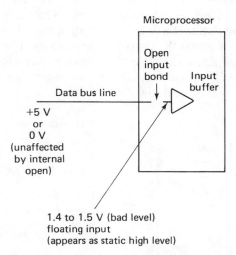

Figure 5-49 Effects of an open input bond.

Short between an input or output and V$_{CC}$ or ground. A short, as shown in Fig. 5-50, has the effect of holding all signal lines connected to that input or output either high (in the case of a short to V$_{CC}$) or low (if shorted to ground). In the case of Fig. 5-50, the address line connected to point A is held in the high (1) state, and the data line connected to point B is held low (0). This will result in a total disruption of the program and is thus one of the easiest types of digital IC failure to locate.

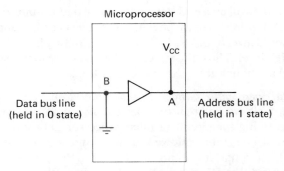

Figure 5-50 Effects of a short between an input or an output and V_{CC} or ground.

Short between two pins. Shorts such as shown in Fig. 5-51 are not as straightforward to analyze as the short to V_{CC} or ground. When two pins are shorted, the outputs driving those pins (the ROM and RAM in this case) oppose each other when one attempts to pull the pins high while the other attempts to pull them low. In this situation, the output attempting to go high will supply current, while the output attempting to go low will sink this current. The net effect is that the short will be pulled to a low state. Whenever both outputs attempt to go high simultaneously or to go low simultaneously, the shorted pins will respond properly. But whenever one output attempts to go low, the short will be held low.

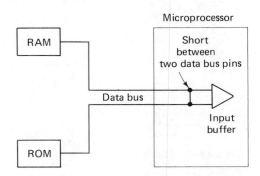

Figure 5-51 Effect of a short between two pins.

Failure of internal control circuitry. The effects of internal failure are difficult to predict. For example, in the case of Fig. 5-52, a failure of the internal circuit has the effect of permanently turning on either the upper transistor of the buffer output totem pole, thus locking the buffer output in the high state or turning on the lower transistor to lock the buffer output in the low state. Of course, this failure blocks data bus signal flow and has a disastrous effect on microprocessor operation.

External shorts. A short between a node and V_{CC} or ground, external to the IC, cannot be distinguished from a short internal to the IC. Both will cause the signal lines connected to the node to be either always high (for shorts to V_{CC}) or always low (for shorts to ground). When this type of failure is found, only a very close physical examination of the circuit will show if the failure is external to the IC. The current tracer (Sec. 5-1-7) is an effective tool for locating such shorts.

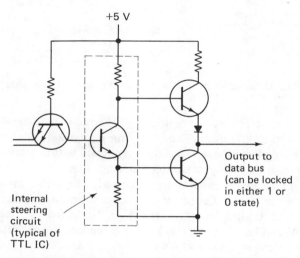

Figure 5-52 Effect of a failure in the internal steering circuitry of a typical TTL IC.

Open signal paths in circuits. An open signal path in the external circuit has an effect similar to that of an open output bond driving the node. As shown in Fig. 5-53, the input (microprocessor) to the right of the open will be allowed to float to a bad level and will thus appear as a static high level in typical TTL circuit operation. Those inputs to the left of the open (data bus) will be unaffected by the open and will thus respond as expected.

5-5 EXAMPLES OF DIGITAL TROUBLESHOOTING

Practical digital troubleshooting is a combination of detective work or logical thinking and step-by-step measurements. Thus far in this chapter, we have described what test equipment is available for digital work and how to

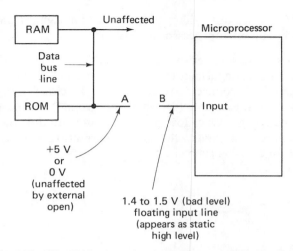

Figure 5-53 Effect of an open signal path in external circuits (data bus).

use the equipment effectively in testing digital circuits, and have given some examples of digital literature. The following paragraphs describe the final step: combining all of these practical techniques with logical thinking to solve some problems in digital troubleshooting.

Two typical digital circuits are discussed, one very simple (the half-adder) and one more complex (a complete decade counter and readout). All three aspects of digital troubleshooting are included: self-check, pulse measurement, and logical thinking. Before going into the specific examples, let us review some very practical details of digital troubleshooting.

5-5-1 Practical digital troubleshooting techniques

The first step in any troubleshooting process is to narrow the malfunctioning area as much as possible by examining the observable characteristics of failure. For example, in the case of a microcomputer, this often involves punching in data at the keyboard and noting the response on the CRT screen. Furthermore, the designers of some microcomputers and other programmed devices have developed *diagnostic routines* or short programs that isolate troubles down to specific ICs or groups of ICs. Such routines should be followed religiously. In the absence of good diagnostic routines and when CRT responses are meaningless, the logic analyzer described in Sec. 5-1-10 is the most effective tool for observing characteristics of failure in programmed digital equipment. As discussed, a logic analyzer makes it possible to observe the full transfer of data in a programmed system (such as a microcomputer) for each step of the program. That is, you can see and compare the data word (on the data bus) at each address (on the address bus) for all program steps.

This can be done one step at a time, in groups of steps, or you can move quickly through the program steps to an area of the program where trouble is suspected.

Based on observations of the logic analyzer (or CRT/keyboard and diagnostic routines), you can localize the failure to as few ICs or other circuits as possible (hopefully to one circuit or IC). At this point it is necessary to narrow the failure further to one suspected circuit by looking for improper key signals or pulses between circuits. The logic probe (Sec. 5-1-5) is a most effective tool for tracing key signals in digital circuits.

In many cases, a signal will completely disappear (no clock signal, no read/write signals, etc.) By rapidly probing the interconnecting signal paths (clock line, read/write lines, etc.), a missing signal can be readily found. Another important failure is the occurrence of a signal on a line that should not have a signal. The pulse memory option found on some probes allows such signal lines to be monitored for single-shot pulses or pulse activity over extended periods of time. The occurrence of a signal will be stored and indicated on the pulse memory LED.

Isolating a failure to a single circuit requires knowledge of the digital equipment circuits and operating characteristics. In this regard, a well-written service manual is invaluable. Properly prepared manuals will show where key signals are to be observed. The logic probe will then provide a rapid means of observing the presence of these signals.

Once a failure has been isolated to a single circuit or IC, the logic probe, logic pulsers, and current tracer (Sec. 5-1) can be used to observe the effects of the failure on circuit operation and to localize the failure to its cause (either an IC or a fault in the circuit external to the IC). The clip and comparator (Sec. 5-1) can also be used, if they will accommodate the number of IC pins.

The classic approach at this point is to test the suspected IC (or ICs) using a logic comparator or by substitution. However, the following steps are based on the assumption that neither of these approaches is practical (the comparator will not accommodate the ICs, and the ICs are very difficult to remove).

Checking for pulse activity on a line. The logic probe can be used to observe signal activity on inputs and to view the resulting output signals. From this information, a decision can be made as to the proper operation of the IC. For example, if a clock signal is occurring on the clock line to a RAM or ROM, and the enabling inputs (such a read or write signals, chip enable signals, etc.) are in the enabled state, there should be signal activity on the data bus lines. Each line should be shifting between high and low (1 or 0) as the program goes through each step. The logic probe will allow the clock and enabling input to be observed and, if pulse activity is indicated on the outputs (each of the data bus lines, in this case), the ROM or RAM can be considered as operating properly. As discussed, it is usually not necessary to measure actual timing of the signals since ICs generally fail catastrophically. A possible exception is when an output buffer does not open and close at the proper time. However, with few exceptions, the occurrence of pulse activity is usually sufficient indication of proper operation.

When more detailed study is desired or when input signal activity is missing, the logic pulser can be used to inject input signals and the probe used to monitor the response. The logic pulser is also valuable for replacing the clock, thus allowing the circuit to be single-stepped, while the logic probe is used to observe the changes in the output state (such as changes on the address and data bus lines).

When using this first step to compare inputs and outputs of the ROM, RAM, and so on, it is wise to check *all the output lines* before you try to check the first fault you find. Prematurely studying a single fault can result in overlooking faults that cause multiple failure, such as shorts between two lines in the data or address bus. This often leads to the needless replacement of a good IC and much wasted time. The extra work can be minimized by systematically eliminating the possible failures of digital circuits, as discussed above.

Testing for an open bond.   The first failure to test for is an open bond in the IC driving the failed node. The logic probe provides a quick and accurate test for this failure. If the output is open, the node will float to a bad level. By probing the node, the logic probe will quickly indicate a bad level. If a bad level is indicated, the IC driving the node is suspect.

Testing for short to V_{CC} or ground.   If the node is not at a bad level, test for a short to V_{CC} or ground. This can be done using the logic pulser and probe. (The current tracer can also be used effectively to find shorts, as discussed in Sec. 5-5-2). While the logic pulser is powerful enough to override even a low-impedance output, the pulser is not powerful enough to effect a change in state on a V_{CC} or ground bus. Thus, if the logic pulser is used to inject a pulse while the logic probe is used simultaneously on the same node to observe the pulse, a short to V_{CC} or ground can be detected. The occurrence of a pulse indicates that the node is not shorted, and the absence of a pulse indicates that the node is shorted to V_{CC} (if the indication remains high) or ground (if the indication remains low).

Cause of V_{CC} or ground shorts.   If the node is shorted to V_{CC} or ground, there are two possible causes. The first is a short in the circuit external to the ICs, and the other is a short internal to one of the ICs attached to the node (as discussed in Sec. 5-4-3). The external short should be detected by an examination of the circuit. If no external short is found, the cause is likely to be any one of the ICs attached to the node. These can be eliminated one at a time. Also check for shorted capacitors and resistors on the line. (Note that resistors do not usually short, but their terminals can be shorted by mechanical vibration.)

Testing for shorts between nodes. If the node is not shorted to V_{CC} or ground, nor is it an open output bond, then look for a short between two nodes. This can be done in one of two ways. First, the logic pulser can be used to pulse the failing node being studied, and the logic probe can be used to observe each of the remaining failing nodes. If a short exists between the node being studied and one of the other failing nodes, the pulser will cause the node being probed to change state (the probe will detect a pulse). To ensure that a short exists, the probe and pulser should be reversed and the test made again. If a pulse is again detected, a short is definitely indicated.

Causes of shorts between nodes. If the failure is a short, there are two possible causes. The most likely cause is a problem in the circuit external to the IC. This can be detected by physically examining the circuit and repairing any solder bridges or loose wire shorts found. Only if the two shorted nodes are common to one IC can the failure be internal to that IC. If after examining the circuit, no short can be found external to the IC, the IC should be replaced.

Testing for an open signal path. If no nodes are indicated as failing, but there is a definite failure of the circuit, an open signal path can be suspected. From a practical standpoint, it is generally easier to check for open signal paths than to replace ICs. The logic probe provides a rapid means of not only detecting but also physically locating an open in a circuit, such as shown in Fig. 5-54. Since an open signal path allows the input to the "right" of the

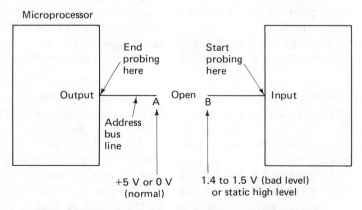

Figure 5-54 Tracing an open signal path in an address bus line.

open (point B) to float to a bad level. Once an input floating at a bad level is detected, the logic probe can be used to follow the circuit back from the input, while looking for the open. This can be done because the circuit to the "left" of the open (point A) will be a good logic level (either high, low, or pulsing), while the circuit to the right (point B) will be a bad level. Probing back along the signal path (from right to left) will indicate a bad level until the open is passed. Thus, the probe will precisely find the physical location of the open.

5-5-2 Digital troubleshooting with the current tracer

The basic procedures for using the current tracer are discussed in Sec. 5-1-7. The following paragraphs describe some additional uses for the current tracer in troubleshooting. From a practical standpoint, whenever a low-impedance fault exists, whether on a digital board or not, the shorted node can be stimulated with a logic pulser and the current followed by means of the current tracer.

Ground planes.   Occasionally, defective ground planes can cause problems in digital circuits. It is possible to determine the effectiveness of a ground plane by tracing current distribution through the plane. This is done by injecting pulse current into the plane from a logic pulser (or any pulse generator) and tracing the current flow over the plane. Generally, current flow should be even over the entire plane. However, it is possible that the current will flow only in a few paths, particularly along the edges of the ground plane.

V_{CC}-to-ground shorts.   The existence of a V_{CC}-to-ground short is generally easy to determine. Typically, there is a drop in power supply voltage or a complete failure of the power supply. However, locating the exact point of the short can be quite difficult. To use the current tracer in this application, disconnect the power supply from the V_{CC} line, and pulse the power supply terminal using the logic pulser with the supply return connected to the ground lead of the pulser. Even if capacitors are connected between V_{CC} and ground, the current tracer will usually show the path carrying the greatest current.

Stuck line caused by dead driver.   Figure 5-55 shows a frequently occurring trouble symptom: a line (say in the address or data bus) has been identified on which the signal is stuck high or low. The particular line is suspect since it shows no pulse activity (as discussed in Sec. 5-5-1). Is the driver (or the signal source) dead, or is something, such as a short, clamping the line to a fixed value (such as to ground or V_{CC})? This question is readily answered by tracing current from the driver to other points on that line. If the driver (source) is dead, the only current indicated by the tracer will be that caused by parasitic coupling from any nearby currents, and this will be much smaller than the normal current capability of the driver. On the other hand, if the

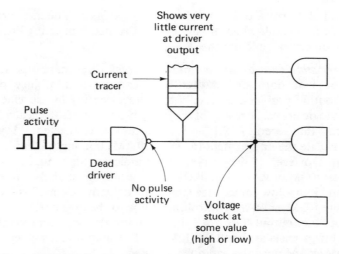

Figure 5-55   Tracing an open signal path in an address or data bus line.

driver is good, normal short-circuit current will be present and can be traced to the short or element clamping the line.

Stuck line caused by input short.   Figure 5-56 illustrates this situation, which has exactly the same voltage symptoms as the previous case of a stuck line caused by a dead driver. However, the current tracer will now indicate a large current flowing from the driver, and will also make it possible to follow

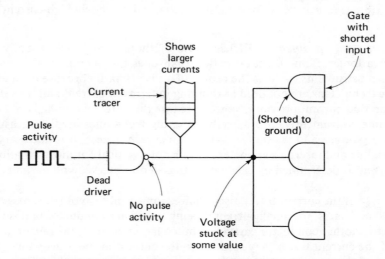

Figure 5-56   Using current tracer to show that stuck node is caused by an input short.

this current to the cause of the problem, a shorted input. The same procedure will also find the fault when the short is on the interconnecting line (for example, a solder bridge to another line).

Stuck three-state data or address bus. A stuck three-state bus, such as a microprocessor data or address bus, can present a very difficult troubleshooting problem, especially to voltage-sensing measurement tools (such as a logic probe). Because of the many bus terminals (typically 8 or 16), and the fact that several ROMs and RAMs may be connected, it is very difficult to isolate the one element (ROM or RAM) holding the bus in a stuck condition. However, if the current tracer indicates high current at several output of a ROM or RAM, it is likely that one (and most likely only one) element is stuck in a low-impedance state. The defective element is located by placing the ROM or RAM control input line to the appropriate level for a high-impedance output state (the "off" condition), and noting whether high current flow persists at the ROM/RAM buffer output. This is repeated for each ROM/RAM until the bad one is located.

If the current tracer indicates high currents at only two elements (say two RAMs), the problem is a *bus fight*. That is, both RAMs are trying to drive the bus at the same time. This is probably caused by improper timing of control signals to the RAMs. (One RAM buffer is opened before the other RAM buffer is closed.)

If the current tracer indicates the absence of abnormally high current activity at all elements, yet the bus signals are known to be incorrect, the problem is an element stuck in the high-impedance state. This can be found by placing a low impedance on the bus, such as a short to ground, and using the current tracer to check for the element that fails to shown high-current activity.

Current tracer problems. Efficient use of the current tracer usually requires a longer familiarization period than does the operation of voltage-sensing instruments (such as the probe). This is primarily because most electronic technicians are not used to thinking in terms of current and the information that current provides, simply because this information has not been available conveniently. Most troubleshooting techniques involve measurement of voltage and resistance. Without a current tracer, it is necessary to open a line and insert an ammeter. This is not practical for most situations, so current is determined by calculation (based on observed voltage and current).

Use of the current tracer also requires some skill to avoid the *cross-talk* problem. That is, if a small current is being traced in a conductor that is very close to another conductor carrying a much larger current, the sensor at the tip of the current tracer may respond to the current in the nearby line. The sensor of the current tracer has been designed to minimize this effect, but cross-talk can never by entirely eliminated. However, by observing the variation of the current tracer's display as the tracer is moved along the line, the operator can learn to recognize interference or cross-talk from a nearby line.

5-5-3 Half-adder troubleshooting

Figure 5-57 shows the diagram and truth table for a half-adder circuit. Assume that this circuit is found in an older computer and is made up of replaceable gates mounted on a plug-in printed circuit card (none of which is typical of present-day equipment, but was quite common a few years ago). The entire half-adder circuit can be replaced as a unit by replacing the plug-in card (as a field-service measure to get the computer operating immediately). Then the card can be repaired by replacing the defective gates (as a factory or shop repair procedure).

Further assume that the computer failed to solve a given mathematical equation during self-check, the trouble is localized to the half-adder card, the card is replaced, and the computer then performs its function normally. This definitely isolates the problem to the half-adder card. Now assume that the card can be connected to a power source (to energize the gates) and that pulses of appropriate amplitude and polarity can be applied to the two inputs (and end and augend). The sum and carry output as well as any other points in the circuit can be monitored with an oscilloscope or logic probe.

Some service shops have special text fixtures that mate with printed-circuit cards. This provides a mount for the cards and ready access to the terminals. In other cases, cards are serviced in the equipment by means of an *extender*. The card is removed from its socket, the extender is installed in the

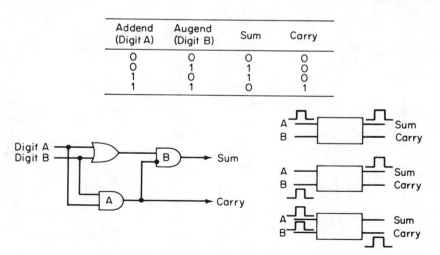

Addend (Digit A)	Augend (Digit B)	Sum	Carry
0	0	0	0
0	1	1	0
1	0	1	0
1	1	0	1

Figure 5-57   Half-adder troubleshooting.

empty socket, and the card is installed on the extender. This maintains normal circuit operation but permits access to the card terminals and components on the cards. These arrangements for printed circuit cards are shown in Fig. 5-58.

To test the half-adder of Fig. 5-57, inject a pulse (true) at the addend input (digit A) and check for a true (pulse) condition at the sum output, as well as a false (no pulse) condition at the carry output. If the response is proper, both the OR gate and the B AND gate are functioning normally. To confirm this, inject a true (pulse) at the augend input (digit B) and check for a true at the sum output, as well as a false condition at the carry output.

If the response is not proper, either the OR gate or the B AND gate are the logical suspects. The A AND gate is probably not at fault, since it requires two inputs (digits A and B) to produce an output. To localize the problem further, inject a pulse at either addend or augend inputs and check for an output from the OR gate. If there is no output or the output is abnormal, the problem is in the OR gate. If the output is normal, the problem is likely in the B AND gate.

To complete the test of the half-adder circuit, inject simultaneous pulses at the addend and augend inputs and check for a true (pulse) at the carry output and a false (no pulse) at the sum output. If the response is proper, all of the gates can be considered to be functioning normally. If the response is not correct, the nature of the response can be analyzed to localize the fault.

For example, if the carry output is false (no matter what the condition of the sum output), the A AND gate produces a true output to the carry

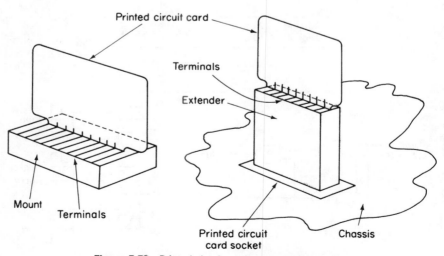

Figure 5-58 Printed circuit card mount and extender.

line when there are two true inputs. Since the test is made by injecting two true inputs, a false condition on the carry line points to a defective A AND gate. A possible exception is where there is a short in the carry line (possibly in the printed wiring). This can be checked by means of the logic probe, pulser or current tracer, as described.

As another example, if the carry output is normal (true) but the sum output is also true, the B AND gate is the most likely suspect. The A AND gate produces a true output when both inputs are true. The B AND gate requires one true input from the OR gate and the false input from the A AND gate to produce a true condition at the sum output (because of the inversion at the input of the B AND gate). If the input to the B AND gate from the A AND gate is true, but the sum output shows a true condition, a defective B AND gate is indicated.

5-5-4 Decade counter and readout troubleshooting

Figure 5-59 shows the logic diagram for a three-digit decade counter and readout. Each digit is displayed by means of a separate seven-segment LED display or readout. Each LED display is driven by a separate BCD seven-segment decoder and storage IC. The three decoder/storage ICs are enabled by a clock pulse at regular intervals or on demand. Each decoder/storage IC receives BCD information from a separate decade counter IC. Each counter IC contains four FFs and produces a BCD output that corresponds to the number of input pulses occurring between reset pulses. The maximum readout possible is 999. At a 1000 count, the output pulse of the 100 decade IC is applied to all three counter ICs simultaneously as a reset pulse.

Troubleshooting problem 1. For the first problem, assume that the pulse input is applied through a gate and that the gate is held open for 1 second. The count shown on the LED readouts then indicates the frequency. For example, if the count is 388, this shows that there are 388 pulses passing in 1 second, and the frequency is 388 Hz.

Assume that a 700-Hz pulse train is applied to the input. The LED display should go from 000 to 700. However, assume that the count is 000, 001, 002, 003, 004, 005, 000, 001, 002, 003, and so on. That is, the count never reaches 006.

One logical conclusion here is that the problem is in the counter function rather than in the readout function (decoder/storage or LED). For example, if the units LED is defective, so that there is no display beyond 005, the 10 and 100 LEDs will be unaffected. That is, the count will be 005, 010, 011, 012, and so on.

With the problem localized to the counter function, there are three logical possibilities: the input pulses never reach more than five (not likely);

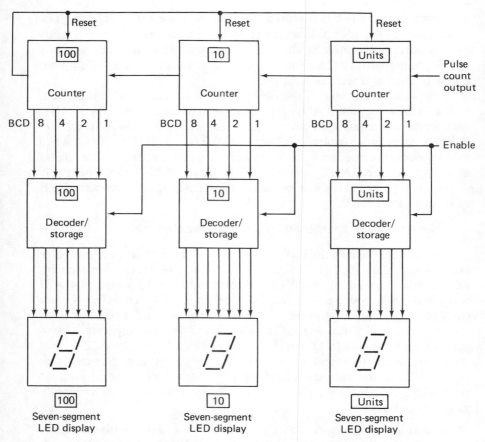

Figure 5-59 Troubleshooting diagram for three-digit decade counter and LED readout.

there is a reset pulse occurring at the same time as the sixth input pulse or any time after the fifth input pulse (possible); and the units counter simply does not count beyond five (most likely).

With the detective work out of my way, the next practical step is to make measurements. To confirm or deny the reset pulse possibility, monitor the input pulse line and the reset line of the units counter on a dual-trace oscilloscope. Adjust the oscilloscope sweep frequency so that about 10 input pulses are displayed.

If a reset pulse does occur after the fifth input pulse, the problem is pinpointed. Trace the reset line to the source of the unwanted (improperly timed) reset pulse. If there is no reset pulse before the sixth pulse, monitor the input line and the 4-line of the units counter. The 4-line should go true at the sixth input pulse (as should the 2-line). Then monitor the 8-line. In all probability, the units counter is defective: the 4- and 2-lines will show no output to the units decoder/storage IC, or the output will be abnormal.

If a logic probe or logic clip is available, check the operation of the units decade counter IC on a single-stepping basis. That is, inject input pulses and check that the 4- and 2-lines go true when the sixth input pulse is injected. If circuit conditions make it possible, disconnect the 4- and 2-lines and recheck operation of the units counter. It is possible that the 4- or 2-line is shorted or otherwise defective. Unfortunately, with most present-day digital circuits, the lines or wiring to the ICs are in the form of printed circuits, making it impractical to disconnect individual lines or leads. The IC package must be checked by substitution (or by comparison if a logic comparator is available for the IC).

Troubleshooting problem 2.   Assume that the test conditions are the same as for Problem 1. However, the readout is now 000, 001, 002, 003, 004, 005, 006, 007, 000, 001, 010, 011, 012, and so on. That is, the 008 and 009 displays are not correct.

The logical conclusion here is that the problem is in the readout function rather than in the counter function. For example, if there is a failure in the counter, the 10 counter will never receive an input from the units counter. Since there is some readout from the 10 counter, it can be assumed that the units counter is functioning.

Assuming that the readout is faulty, there are several logical possibilities: the units counter output lines can be shorted or broken (thus, the units decoder receives no input or an abnormal input); the units decoder output lines can be shorted or broken (thus the units LED receives an abnormal input); or the units LED is defective.

The first practical step depends on the test equipment available. Monitor the units decoder output to the LED segments. (The truth table for a seven-segment readout is given in Fig. 3-41.) If pulses are present on all segments but there is no 8 or 9 display, the LED is at fault. As an alternate first step, inject a pulse at all segments of the LED, and check for an 8 display.

If the pulses required to make an 8 or 9 display are absent from the LED, check for an 8 or 9 input to the units decoder. An 8 input is produced when there is a pulse at the BCD 8-line (between counter and decoder). A 9 input to the decoder requires simultaneous pulses on the 8- and 1-lines.

If the 8 and 9 inputs are available to the decoder but there are no 8 and 9 pulses to the LED segments, the units decoder is at fault. (As shown in Fig. 3-41, an 8 display requires that all segments receive pulses; a 9 display requires all segments but the *d* and *e* segments.) If a logic clip is available, check the operation of the units decoder on a single-stepping basis. That is, inject input pulses at the BCD 8-line (from the counter) while simultaneously enabling the decoder, and check that the outputs to all LED segments go true.

Troubleshooting problem 3.   Assume that the test conditions are the same as for Problem 1 except that the input frequency is 300 Hz. However, the readout is 600. That is, the readout is twice the correct value. The logical conclusion here is that the problem is in the counter function rather than in the readout. It is possible that the readout could be at fault, but it is not likely.

A common cause for faults of this type (where FFs are involved) is that one FF is following the input pulses directly. That is, the normal FF function is to go through a complete change of states for two input pulses. Instead, the faulty FF is changing states completely for each input pulse (similar to the operation of a one-shot multivibrator). Thus, the decade counter containing such a faulty FF produces two output pulses to the next counter for every 10 input pulses, or the counter divides by 5 instead of 10. All decades following the defective stage receive two input pulses, whereas they should receive one.

The first practical step is to monitor the input and output of each decade in turn, starting with the units decade. The decade that shows two outputs for 10 inputs, or one output for 5 inputs, is at fault.

Troubleshooting problem 4.   Assume that the test conditions are the same as for Problem 1. However, the readout is now 000, 000, 000, 000, and so on. That is, the LEDs glow but remain at 000.

The logical conclusion here is that the LEDs are receiving power and are operative. If not, the LEDs could not produce a 000 indication. The most likely causes for such a symptom are: (1) no input pulses arriving at the units counter; (2) a simultaneous reset pulse with the first input pulse (or a short on the reset line); (3) no enable pulse to the decoder/storage ICs; or (4) a defective units counter (not responding to the first input pulse).

Here, the practical steps are to monitor (simultaneously) the input line and reset line, input line and enable line, and input line and the output of the units counter IC. This should pinpoint the problem. For example, if there are no input pulses (or the wiring is defective, possibly shorted, so that the input pulses never reach the units counter input), there will be no output. If the input pulses are present but a reset pulse arrives simultaneously with the first input pulse, there can be no readout.

If there is no enable pulse, there will be no readout, even with the counters operating properly. That is, the counters will produce the correct BCD output, which is then stored in the decoders. However, the absence of an enable pulse will prevent the stored information from being displayed on the LEDs. If the input pulses are present and there is no abnormal reset pulse, the output of the units counter should show one pulse for 10 input pulses. If not, the units counter is defective.

Index